The Emperor Constantine

IN THE SAME SERIES

General Editors: Eric J. Evans and P.D. King

LANCASTER PAMPHLETS

The Emperor Constantine

Hans A. Pohlsander

London and New York

First published 1996
by Routledge
11 New Fetter Lane, London EC4P 4EE

Simultaneously published in the USA and Canada
by Routledge
29 West 35th Street, New York, NY 10001

Typeset in Bembo by
Ponting–Green Publishing Services, Chesham, Bucks
Printed and bound in Great Britain by
Clays Ltd, St Ives PLC

British Library Cataloguing in Publication Data
A catalogue record for this book is available from
the British Library

Library of Congress Cataloguing in Publication Data
Pohlsander, Hans A.
Constantine / Hans A. Pohlsander.
(Lancaster pamphlets)
Includes bibliographical references.
1. Constantine I, Emperor of Rome, d. 337.
2. Emperors–Rome–Biography
3. Rome–History–Constantine I, the Great, 306–336.
I. Title. II. Series.
DG315.P65   1996
937'.08–dc20   96–10186

ISBN 0–415–13178–2

Contents

Appendices

Figures

Foreword

Lancaster Pamphlets offer concise and up-to-date accounts of major historical topics, primarily for the help of students preparing for Advanced Level examinations, though they should also be of value to those pursuing introductory courses in universities and other institutions of higher education. Without being all-embracing, their aims are to bring some of the central themes or problems confronting students and teachers into sharper focus than the textbook writer can hope to do; to provide the reader with some of the results of recent research which the textbook may not embody; and to stimulate thought about the whole interpretation of the topic under discussion.

Acknowledgements

My thanks are due to Professor Louis W. Roberts, my colleague at the University at Albany, who has read parts of my manuscript and provided valuable insights, to Professor Charles Odahl, of Boise State University, who has read the entire manuscript and made numerous suggestions for improvement of my text, and to Professor Eric J. Evans of Lancaster University, who has guided this project from its inception to its completion.

TIMECHART

Chief dates of Roman history, 235–337

310	Death of Maximian
311	Galerius' "edict of limited toleration"
311, 11 May	Death of Galerius
312, 28 Oct.	Battle at the Milvian Bridge
313, (?) Feb.	Meeting of Constantine and Licinius at Milan; marriage of Licinius to Constantia, half-sister of Constantine
313	"Edict of Milan"; Donatist controversy
314	Council of Arles
315	Constantine observes his *decennalia*
316	First war between Constantine and Licinius
317, 1 March	Crispus, Constantine II, and Licinius II appointed Caesars
324	Second war between Constantine and Licinius
324, 8 Nov.	Constantius II appointed Caesar
325, Spring	Death of Licinius I and Licinius II
325, 20 May– 26 July	Council of Nicaea; adoption of the Nicene Creed
325–6	Constantine observes his *vicennalia*
326	Death of Crispus and Fausta; pilgrimage of Helena to the Holy Land
330, 11 May	Dedication of Constantinople
333, 25 Dec.	Constans appointed Caesar
335–6	Constantine observes his *tricennalia*; Council of Tyre
335, 13 Sep.	Church of the Holy Sepulchre in Jerusalem dedicated
335, 7 Nov.	Athanasius exiled to Trier
337, after Easter	Constantine baptized by Bishop Eusebius of Nicomedia
337, 22 May	Death of Constantine at Nicomedia
337	Army coup in Constantinople
337, 9 Sep.	Constantine II, Constantius II, and Constans assume the title of Augustus

1
Introduction

The emperor Constantine has been called the most important emperor of Late Antiquity. His powerful personality laid the foundations of post-classical European civilization; his reign was eventful and highly dramatic. His victory at the Milvian Bridge counts among the most decisive moments in world history.

But Constantine is also controversial, and the controversy begins in antiquity itself. Julian the Apostate accused Constantine of greed and waste. The pagan historian Zosimus held Constantine responsible for the collapse of the (Western) empire. The Christians Lactantius and Eusebius, on the other hand, saw in him a divinely appointed benefactor of mankind. This positive view prevailed throughout the Middle Ages.

In more modern times Constantine has come in for some harsh criticism. Thus Edward Gibbon, in his celebrated *The Decline and Fall of the Roman Empire* (1776–88), held that Constantine degenerated "into a cruel and dissolute monarch," one who "could sacrifice, without reluctance, the laws of justice and the feelings of nature to the dictates either of his passions or of his interest." He also held that Constantine was indifferent to religion and that his Christian policy was motivated by purely political considerations.

In his *The Age of Constantine the Great* (1852) the renowned Swiss historian Jacob Burckhardt saw in Constantine an essentially unreligious person, one entirely consumed by his ambition and lust for power; worse yet, a "murderous egoist" and a habitual breaker of oaths. And, according to Burckhardt, this man was in matters of religion not only inconsistent but "intentionally illogical."

In our own century competent historians of antiquity have examined the record more objectively and reached more balanced conclusions. In the following chapters an effort will be made to present these more balanced conclusions in a concise form, or, where conclusions are not possible, the issues and problems in an unbiased manner.

2
The soldier emperors and Diocletian

If we wish to understand the emperor Constantine we must first examine briefly the times in which he was born and raised and which left their mark upon him.

The half century which passed from the death of the emperor Severus Alexander in 235 to the accession of the emperor Diocletian in 284 presented the Roman empire with a seemingly endless series of crises and calamities, political, military, economic, and social in nature.

One clear indication of the insecurity of the times is found in the rapid succession of emperors. With predictable regularity emperor after emperor rose to the top through the ranks of the army, reigned for a short while, and died on the field of battle or fell victim to assassination. The average length of the reigns of these emperors is only three years, and none lasted for more than eight years. How many emperors there were is difficult to say with any degree of accuracy, because in addition to those who gained recognition by the senate there were numerous usurpers and contenders. All of them had risen through the ranks of the army and are, therefore, often called the soldier emperors or barrack-room emperors. Many of them were capable and energetic enough, but none was able to break the vicious cycle. At the same time the integrity of the empire was threatened by separatist movements in both West and East: the emperor Aurelian (270–5) had to overcome both a secessionist

3

Palmyrene kingdom, under the famous Zenobia, in the East and a secessionist Gallo-Roman empire in the West.

Along the long Rhine-Danube frontier the Romans faced larger, better organized and more formidable Germanic tribes than ever before: Saxons, Franks, Alamanni, Marcomanni, Vandals, Burgundians, Visigoths, and Ostrogoths. Again and again one or the other of these tribes penetrated deep into Roman territory. Gaul and northern Italy suffered especially from repeated Germanic incursions; Dacia (modern-day Romania) had to be abandoned. Even the safety of the city of Rome could not be taken for granted; Aurelian's Wall, 12 miles long and still almost encircling the heart of the city, is a monument not only to Roman engineering skills but also to Roman insecurity. In the East Sassanid Persia pursued a vigorous policy of aggression and expansion. In 260 the emperor Valerian was taken prisoner by the Persian king Shapur I at Syrian Edessa and suffered unspeakable humiliation before dying in captivity.

In circumstances such as these, it is easy to see that the need to recruit, pay and supply a large standing army took precedence over all other needs. Taxation and requisition of goods, often unfairly and inefficiently administered, imposed intolerable burdens on the civilian population and weakened the entire economy. Inflation ran rampant, and the currency was debased. Agricultural production declined as the rural population was driven by despair to abandon productive land and take flight. There was an increase in brigandage. In the towns it became increasingly difficult to get the *curiales*, members of the propertied class, to fill administrative posts and to assume the financial burdens which went with these posts. To add to all these woes, parts of the empire, notably North Africa and the Balkans, were visited by the plague.

It seems incongruous to the modern observer that, while the empire was thus troubled, the emperor Philip the Arab (244–9) should in 248 observe the one-thousandth anniversary of the founding of the city of Rome with extravagant and wasteful Secular Games. But the prevailing sentiment of the age was a deeply conservative one: salvation was sought not in daring innovation but in a return to traditional practices, institutions and values. It is in this context also that we must understand the anti-Christian measures ordered by the emperor Decius in

4

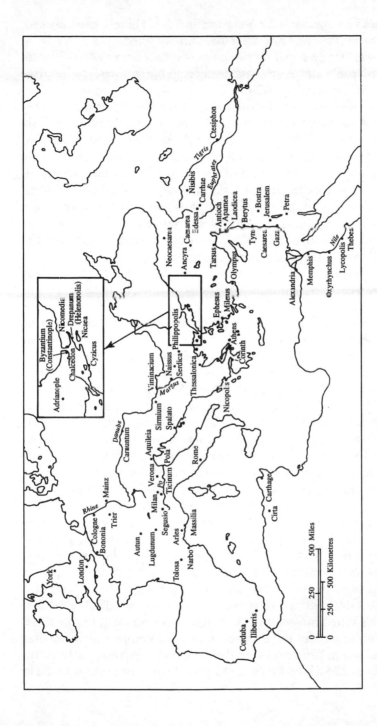

Figure 1 Map of the Roman empire.

Courtesy of the Cartography Office, Department of Geography, University of Toronto.

249 and by the emperor Valerian in 257. These measures, too, fell short of success and were soon abandoned.

It was finally given to the emperor Diocletian (284–305) to restore the empire, by his indefatigable efforts, to relative security and stability. Who was this remarkable man?

Diocletian, born under the name of Diocles, was a man of very humble background, as many of the soldier emperors before him had been. No longer did a humble background keep a career officer from rising to the highest ranks, ranks once the preserve of the senatorial class. And like several of the soldier emperors before him, like Decius or Claudius Gothicus for instance, he was a native of Illyricum (the Balkans). This is not a co-incidence, because this region was known for its adherence to traditional Roman values, such as patriotism, discipline and piety, and for the quality of its recruits.

Diocles, the future Diocletian, was born on 22 December in 243, 244, or 245, in Salona, today a suburb of Split, on the Dalmatian coast. His father was a freedman, or perhaps he himself was a freedman. He received only a limited education and, like so many of his countrymen, sought a career in the army. We do not know the details of this career before 284.

Numerian, son of the short-lived (282–3) emperor Carus, was from July 283 to November 284 joint emperor with his older brother Carinus; Numerian was stationed in the East and Carinus in the West. Numerian, by reason both of his youth and of his temperament, was ill-suited for the demands made upon him. It was the praetorian prefect Arrius Aper who actually controlled things, and it is he who killed Numerian at Nicomedia in the hope of succeeding to the throne himself; this Aper must have been not only a powerful but also a ruthless man, for the young man whom he dispatched was his own son-in-law. Aper's designs did not lead to the desired end: a council of the army, meeting at Nicomedia on 20 November 284, proclaimed as the next emperor not Aper, but Diocles, then the commander of the *protectores domestici*, the "household cavalry." As for the hapless Aper, the soldiers seized him, and Diocles killed him on the spot. But there was still the problem of Numerian's older brother Carinus, whom the army had passed over. The armies of the two rivals met in battle in the spring of 285 at the Margus (Morava) River near Belgrade. The

6

forces of Carinus routed those of Diocletian, but Carinus was killed in action, making the victory pointless; both armies declared for Diocles (Diocletian).

For the Roman empire of the third century the manner in which the new emperor had risen to the very top was the norm rather than the exception. What he had done had been done many times before; what made him different from his predecessors was that he succeeded where they had failed and that he endured for more than twenty years while they had been swept away after much shorter reigns. He broke the cycle. Nor can much moral blame be attached to what he did; any other course of action might well have cost him his own life.

The new emperor took on the name of Gaius Aurelius Valerius Diocletianus and began the daunting task of restoring the empire. He, too, like the men before him, was conservative in his approach: traditional institutions were to be restored or strengthened, not replaced, and in this conservative approach religion played a major role. He was capable, confident and proud, and as a soldier he knew the value of discipline.

Early in his reign, certainly before the end of 285, Diocletian appointed a trusted comrade and fellow-Illyrian, Maximian, as Caesar and gave him responsibility for the western part of the empire, especially for the security of the Rhine frontier. Before too long, on 1 April 286, he promoted him to the rank of Augustus. Joint rule of two Augusti had been tried before, but less successfully: Marcus Aurelius had in the earlier years of his reign as his fellow-Augustus the undistinguished Lucius Verus, and Gallienus had been Augustus jointly with his father Valerian. On 1 March 293 Diocletian expanded this joint rule of two Augusti into the system now known as the Tetrarchy, or First Tetrarchy, to distinguish it from the Second Tetrarchy which existed for a short while after his retirement in 305. He did this by appointing two other Illyrians, Galerius and Constantius, as Caesars or "junior emperors." Galerius was to serve as Caesar under Diocletian himself, and Constantius, the father of Constantine, was to serve as Caesar under Maximian. To strengthen the bonds between each Caesar and his Augustus, both Caesars were adopted by their respective Augusti; furthermore Valeria, Diocletian's daughter, was given in marriage to Galerius, and Theodora, Maximian's daughter (or stepdaughter, according to some sources), to Constantius.

7

The new system did not divide the empire. Each of the four emperors had specific responsibility, especially in matters of defence, but they were not restricted by territorial boundaries. Each Augustus supervised his own Caesar, and Diocletian reserved senior authority for himself. All new legislation was issued in the name of both Augusti. The new system provided for greater efficiency in the administration of the empire and for greater security along the borders (by reducing response time). It also provided an orderly means of succession: it was anticipated that in due time the two Augusti would step down, the two Caesars would succeed them, bringing with them a wealth of experience, and they in turn would appoint new Caesars. The system was thus, in theory, self-perpetuating; it would prevent the disgraceful way in which most emperors during the third century, including Diocletian himself, had attained their position. That it worked only once certainly was not Diocletian's fault.

Diocletian provided for a new territorial organization of the empire as well. There were now four areas of responsibility, the later prefectures: the West, Italy, Illyricum and the East. Each of these had its own capital or, more accurately, chief imperial residence: Constantius resided in Trier, Maximian in Milan, Galerius in Thessalonike, and Diocletian in Nicomedia (now Izmit). The city of Rome, although still the seat of the senate, was now of much diminished importance. The second-highest level of administration was the diocese, and there were twelve dioceses in the empire, each headed by a *vicarius* and each consisting of a number of provinces. By dividing existing provinces Diocletian vastly increased the number of provinces throughout the empire from about forty to more than one hundred; it has often been assumed that the aim was to reduce the possibility of rebellion. It is also noteworthy that Italy, except for the city of Rome, lost its privileged status and was divided into provinces and subjected to taxation just as the other territories of the empire were.

Diocletian and his colleagues were quite successful in their efforts to assure the integrity and security of the empire. Diocletian himself campaigned successfully on the Danube frontier and settled a revolt in Egypt. His Caesar Galerius fought successfully against the Goths on the lower Danube and, after an initial setback, won a fine victory in 298 over the Persian

8

king Narses; the Arch of Galerius in Thessalonike bears witness of that victory to this day. Maximian secured the Danube frontier and quelled an insurrection in Mauretania. His Caesar Constantius secured the Rhine frontier and recovered Britain from the secessionist naval commander Carausius and from the usurper Allectus who succeeded him. To achieve these military successes Diocletian had vastly increased the size of the army, perhaps doubled it. He had also divided it into two branches, the mobile field army, known as the *comitatenses*, and the border garrison troops, known as the *limitanei*.

Diocletian was much less successful in his attempts to strengthen the faltering Roman economy. The burdens placed upon this economy were now heavier than ever because of the increased size of the army, an enlarged bureaucracy, and four imperial residences. An equalization of the tax assessments provided at least some relief. Diocletian's monetary reform, undertaken in 294, was only partially successsful; a currency edict in 301 represents a further effort to stabilize the monetary system. His famous Edict on Prices, also issued in 301, was an attempt to curb inflation by prescribing maximum prices for a long list of goods and services. It failed, but to historians the extant text is the single most important document of Roman economic history. In his efforts to control the economy Diocletian even used coercive measures to freeze people in their occupations, whether they be farmers or craftsmen, and in their habitats. He was, of course, no more an economist than anyone else in antiquity.

Diocletian's desire for conformity, including religious conformity, and also political considerations led him in 302 to launch a severe persecution of the Manichees, who were followers of the Mesopotamian visionary Mani (214–76) and perceived as friends of Persia. But Manichaeism survived the persecution, and later in the fourth century St Augustine was among its followers for some nine years.

The Christians' turn was to come a little later, although it was reported that Diocletian's wife Prisca and his daughter Valeria were pro-Christian. On 23 February 303 the Christian church in Nicomedia was, on Diocletian's orders, torn down. The first edict of persecution was issued on the following day. It did not come as a total surprise; some time earlier there had been an embarrassing incident at the auguries, and pagan priests had

9

blamed the presence of Christians; also Diocletian had already begun to dismiss Christians from the army and the imperial service. This first edict ordered the Christians to turn over their scriptures to be burned; it also ordered all Christian churches to be destroyed. A second edict followed in the summer, after two suspicious fires in the imperial palace in Nicomedia; it ordered all Christian bishops to be arrested. The third edict, later in the same year, directed that the imprisoned Christians be forced to make sacrifice and then be released; the prisons, perhaps, were filled to capacity. In the midst of these events Diocletian proceeded to Rome, there to celebrate, together with Maximian, the beginning of the twentieth year of his reign, his *vicennalia*; this was his first and only visit to the city of Rome. A fourth edict, probably in the spring of 304, finally ordered all Christians everywhere to make sacrifice or face execution. Enforcement of these edicts was uneven throughout the empire; while it was severe in the East, it was patchy elsewhere and particularly lax in the territories governed by Constantius. Although this persecution lasted for years it achieved no lasting results.

In his *On the Deaths of the Persecutors* Lactantius places the blame for the persecution squarely on the shoulders of Galerius and, in keeping with the purpose of this pamphlet, sees in Galerius' painful death, described in lurid detail, the will of God. Eusebius first, in his *History of the Church*, expresses the opinion that with the persecution God meant to chastise the Christians for their internal dissension; later, in his *Life of Constantine*, he, too, holds Galerius responsible. And Constantine, in his *Oration to the Assembly of the Saints* (if correctly interpreted), also names Galerius as the author of the persecution. Explanations offered by modern scholars usually endeavor to see beyond the hatred for the Christians which Lactantius attributes to Galerius; Diocletian's desire to foster traditional Roman values certainly was a factor.

Diocletian and Maximian had claimed for themselves the titles of Jovius and Herculius respectively, thus not only reasserting the ancestral religion, but also suggesting to their subjects that they enjoyed divine patronage. Increasingly surrounding themselves with the symbolism and ritual of divinely sanctioned imperial power, they and everything pertaining to them became sacred. But the gods granted them partial success only; both the Edict on Prices and the Great Persecution were failures. Historians recognize Diocletian's superb organizational talent; he has even been called the most remarkable imperial organizer since Augustus and a statesman of the first rank. Without question he put an end to anarchy and restored stability, against overwhelming odds. He also profoundly changed the nature of Roman emperorship: the emperor was no longer a *princeps*, a "first man" who shared or pretended to share authority with the senate, but rather a *dominus*, a "lord" who ruled with absolute power.

On 1 May 305 Diocletian and a reluctant Maximian announced their retirement.

3

Constantine's rise to power

The future emperor Constantine was born to the future emperor
Constantius by a woman named Helena. It is expedient now to
relate what is known about the background of this remarkable
woman and about her relationship to Constantius.

Helena's birthplace, in all likelihood, was the town of
Drepanum in Bithynia (in northwestern Asia Minor), which
Constantine later renamed Helenopolis in his mother's honor.
The late and highly unlikely claims of other cities in Europe and
Asia, such as Colchester in England, Trier in Germany, and
Edessa in Syria, must be rejected. The date of Helena's birth can
be reconstructed only from what we know about her death. We
know from reliable sources that she undertook a pilgrimage to
the Holy Land after the deaths of her grandson Crispus and her
daughter-in-law Fausta in 326 and that she died not too long
after her return from that pilgrimage at the age of eighty. Thus
we may conclude that she died in c. 328–30 – numismatical
evidence points more specifically to the year 329 – and that she
was born in 250 or shortly before.

In the course of recording Constantine's parentage a number
of ancient authors tell us that Helena was Constantius' concu-
bine, while others do not specify her legal or social status and
yet others call her Constantius' *uxor* (wife). The legal status of
Helena's and Constantius' relationship thus remains subject to
question, but most modern scholars do not think that Helena

was legally married to Constantius. It is apparent that Helena was of low social status. St Ambrose of Milan reports that she was a *stabularia*, a maid in a tavern or inn; a good *stabularia*, he adds. The inn in which Helena was employed, according to a late source, was in her native Drepanum and was operated by her parents, and here she is said first to have met Constantius. After their first meeting, wherever and whenever that may have taken place, Helena must have accompanied Constantius on his military assignments, for she gave birth to Constantine at Naissus in the province of Moesia Superior, the modern Nish in Serbia.

Reliable sources provide us with the month and day of Constantine's birth, namely 27 February, but deny us the year. A number of remarks made by Eusebius in his *Life of Constantine* allow us to conclude that it occurred between 271 and 273; some other sources place it variously between 272 and 277. Attempts through the years by a number of scholars to place Constantine's birth in the 280s have been cogently refuted more recently.

Constantius' career took him to the very top. In 271–2 he served under Aurelian in the East as a member of the *protectores* (officers in attendance on the emperor), and later he attained the rank of tribune. In 284–5 he was the *praeses* (a provincial governor lower in rank than a *consularis*) of Dalmatia. In all likelihood he was Maximian's praetorian prefect in 288–93. On 1 March 293, as we have already seen, he was raised to the rank of Caesar. Several ancient sources record that *on this occasion* he was required to put aside Helena and to marry Theodora, the daughter of Maximian. On the other hand a panegyric addressed to Maximian on 21 April 289 contains a remark which has been interpreted by numerous scholars as indicating that Constantius was already married to Theodora at that time. The present writer has been persuaded by the evidence that the separation of Helena and Constantius took place in 293, as a precondition of Constantius' appointment, and not at an earlier date. We do not know where or under what circumstances Helena lived after her separation from Constantius, and nothing suggests that their paths ever crossed again. Nor do we know in what year or at what age Constantine was separated from his mother or whether they had any opportunity to see each other during the long years of separation. That there were strong

13

bonds of affection between mother and son will become clear as the story unfolds further.

In 305, before announcing his retirement, Diocletian conferred at Nicomedia with his Caesar Galerius on the choice of two new Caesars; Maximian and Constantius apparently were not consulted. The choice fell on Severus, an army officer from Pannonia (Hungary), and on Maximinus Daia, or Daza, another army officer, who was Galerius' nephew and, like Galerius himself, fiercely anti-Christian. Maxentius, son of Maximian, and Constantine, son of Constantius, were pointedly passed over. The former, although married to Galerius' daughter Valeria, was undistinguished. But the latter had served with distinction in the East both under Diocletian and under Galerius, had risen to the rank of tribune, and enjoyed great popularity with the soldiers; his elevation to imperial rank was widely anticipated.

On 1 May 305 Diocletian conferred the purple on Maximinus Daia at Nicomedia in front of the assembled army, and on the same date Maximian conferred the purple on Severus at Milan in a similar setting. Galerius and Constantius advanced to the rank of Augustus, replacing the retirees Diocletian and Maximian, who now became private citizens. Diocletian withdrew to the palace which he had built for himself at Salona, while Maximian withdrew to an estate in Lucania (in southern Italy).

The members of the Second Tetrarchy divided the provinces of the empire among themselves, just as the members of the First Tetrarchy had done. Constantius retained Gaul and Britain and gained Spain. Italy, Africa and Pannonia were assigned to Severus. Galerius chose the Balkans, excepting Pannonia, and Asia Minor. The eastern provinces were assigned to Maximinus Daia.

There was a good deal of tension among the four emperors, now that the restraining influence of Diocletian was removed, and especially between Constantius and Galerius. Constantius requested that his son Constantine, who had been with Diocletian and Galerius in the East, be allowed to join him. Galerius consented, and Constantine rushed to his father's side, not giving Galerius a chance to change his mind. Constantine met his father at Gesoriacum (Boulogne), crossed the Channel with him, and assisted him in a campaign against the Picts (the people beyond Hadrian's Wall in modern-day Scotland). Then, on

14

25 July 306, Constantius died at Eburacum (York), and the soldiers at once proclaimed Constantine their Augustus, doing what Roman armies had done so often in the past. The dynastic principle had proven stronger than the tetrarchic principle. Constantine henceforth observed 25 July as his *dies imperii*, thus sanctioning the soldiers' illegal action. Galerius, upon learning what had happened, offered a compromise: the rank of Augustus belonged by rights to Severus, but Constantine could be Caesar. Constantine, for the moment, acquiesced. But peace and the Third Tetrarchy did not prevail for long.

In Rome the Senate and the Praetorian Guard had long resented the loss of power, prestige, and privileges. They found an ally in Maximian's son, Maxentius, and on 28 October 306 proclaimed him emperor, initially in the rank of *princeps*. Then Maximian returned from the retirement which he had only reluctantly accepted, not only to support his son, but also to claim again for himself the powers of an Augustus. Severus was unable to oppose Maxentius and Maximian effectively. He had not been popular in Italy, and his army was weakened by desertions. He withdrew to Ravenna, where, in the spring of 307, he surrendered to Maximian, who had promised to spare his life but soon forced him to commit suicide.

In the meantime Maximian and Constantine had reached an understanding with each other, and Constantine did not come to the rescue of the beleaguered Augustus Severus. In 307 Maximian and Constantine jointly claimed the consulship, but were recognized only in their own domains. In September of 307, at Trier, Constantine married Maximian's daughter Fausta, putting away his mistress Minervina, who had borne him his first son, Crispus. When Maximian and Maxentius fell out with each other Maximian sought refuge with Constantine in Gaul. Attempts by Galerius to dislodge Maxentius failed, and the latter, on 27 October 307, claimed the rank of Augustus.

Diocletian briefly emerged from retirement to attend a conference with Maximian and Galerius on 11 November 308, at Carnuntum (near Vienna), as we know not only from literary sources but also from a votive inscription to Mithras left by the three emperors. He was invited to reclaim power as senior Augustus, but he declined; he derived more pleasure, he declared, from raising cabbage in the garden of his palace. He never became involved again. The date of his death is not known

15

with certainty; 311, 312 and 313 have been proposed. His palace, containing his mausoleum, stands to this day as one of the greatest architectural achievements of Late Antiquity.

An agreement was hammered out among the three emperors attending the meeting at Carnuntum: Maximian was to return to retirement, and Maxentius was declared a usurper (as was a certain Domitius Alexander, who had seized Africa and controlled it until he was overcome by Maxentius in 311). In an attempt to preserve the tetrarchic principle, Galerius appointed a new Augustus, Licinius. This appointment slighted the two men already serving as Caesars, Maximinus Daia and Constantine; they, too, became Augusti before long, not being content with the title *filius Augusti* offered to them by Galerius.

While Constantine had provided Maximian with a place of refuge and had married his daughter Fausta, he had been careful, nevertheless, not to grant the old man any power or authority. Maximian abused the hospitality shown him, ignored the bonds of kinship, and in 310 attempted one more grab for power. While Constantine was campaigning against the barbarians on the Rhine, he seized the treasury at Arles and proclaimed himself Augustus once again. The attempt failed; Constantine moved against him swiftly and resolutely. Maximian had to surrender at Massilia (Marseilles) and was soon forced to commit suicide, after one last futile plot on Constantine's life, according to a lurid tale told by Lactantius.

That left five contenders on the field: Constantine, Maxentius, Galerius, Licinius, and Maximinus Daia. Galerius was to be eliminated first. In May 311 he succumbed to a horrible disease (cancer of the bowels?) at Nicomedia; he was buried in his native Romulianum on the Danube, not in the splendid mausoleum which he had built for himself at Thessalonike. The surviving four emperors, all in the rank of Augustus, were deeply suspicious one of another.

In these turbulent years Constantine was occupied not only with asserting himself against actual and potential challengers, but also with defending his realm against the barbarians along his frontiers. After he had been proclaimed emperor at York, on 25 July 306, he settled affairs in Britain quickly and returned to the Continent. For the next six years Trier was to serve him as his principal residence, as it had served his father Constantius before him, and before that the emperor Maximian. He stayed in the city again for extended periods of time in 313, 314, 316 and, for the last time, in 328. Using Trier as his base of operations, he campaigned successfully against the Franks in 306–7 and against the Bructeri (north of the Ruhr) in 307–8. Two Frankish kings captured in the course of the former campaign he fed to the beasts in the amphitheatre of Trier; in the course of the latter campaign he constructed a bridge across the Rhine at Colonia Agrippina (Cologne). He was campaigning against the Franks and Alamanni in 310 when he received word of Maximian's usurpation. He also found time for two visits to Britain, one in 307, the other probably in 310.

At Trier, too, Constantine was joined by his mother Helena, for whom it was at last safe to emerge from the obscurity which had been her lot since her separation from Constantius. Today any visitor, even one with only a minimal knowledge of or interest in Roman history, can still see the evidence of the city's former imperial splendor. Trier's most famous landmark, the *Porta Nigra*, dates from the second century, as do the Roman bridge across the Mosel and the Baths of St Barbara; the amphitheatre was constructed even earlier, at the end of the first century. But the Imperial Baths (Figure 2), impressive even in their ruinous state, and the so-called Basilica, actually the reception hall (*aula palatina*) of the imperial palace, are both to be associated with Constantine.

Trier's cathedral, dedicated to St Peter, has a long history which ultimately reaches back to the time of Constantine. Our mediaeval sources report that Helena donated her "house" in Trier, so that it might become a church, and that Bishop Agritius of Trier dedicated this church. The mediaeval sources which offer this statement are generally not very reliable, and it is known that the cathedral was not completed until many years after Helena had left Trier and also some years after Bishop Agritius had died. It is not surprising, therefore, that in the past scholars have denied that Helena had anything to do with the beginnings of Trier's cathedral. But archaeological excavations underneath the nave of the cathedral in 1945–6 and again in 1965–8 have dramatically changed that view. These excavations revealed a room, measuring *c.* 10 × 7 m, which had been built after 316, but was torn down after 330. The location, the date of construction and the quality of the decorations have persuaded investigators that this room was once part of the

17

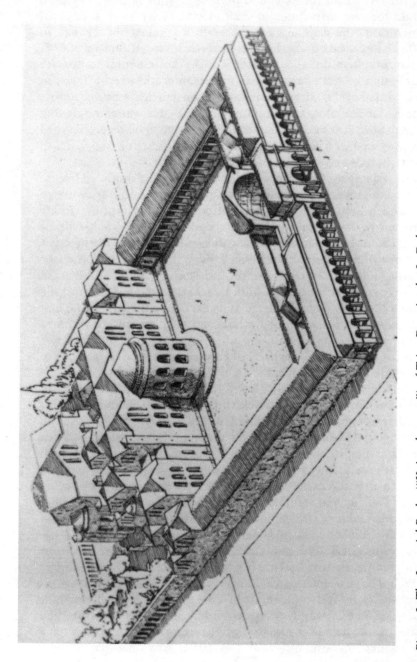

Figure 2 The Imperial Baths ("Kaiserthermen") of Trier. Drawing by L. Dahm. Courtesy of Rheinisches Landesmuseum, Trier.

imperial residence, "Helena's house," and that it was taken down to make room for the Constantinian church. The painted ceiling of the room was found *in situ* but in thousands of fragments. The fifteen panels which made up this ceiling have been painstakingly restored and are exhibited in Trier's diocesan museum. Four of these fifteen panels depict richly dressed and bejewelled ladies, who represent either members of the imperial family or allegorical figures. If the former interpretation be right, the four ladies would be Constantine's mother Helena, his wife Fausta, his half-sister Constantia, and the wife of his son Crispus, also named Helena. This writer is persuaded that the latter interpretation is the correct one.

Although Constantine and Maxentius were brothers-in-law, relations between them were strained. Maxentius and Maximian had quarrelled, as we have seen, and Maximian had sought refuge with Constantine. But after Maximian's death at Constantine's hands Maxentius professed to avenge his father and prevailed on an obedient senate to deify him. When Constantine betrothed his half-sister Constantia to Licinius, late in 311 or early in 312, Maxentius felt threatened. Before long Constantine's statues in Rome and elsewhere in Italy were thrown down. The message was clear, and Constantine was not slow to respond. In the summer of 312 he crossed the Graian Alps by way of the Mt Cenis Pass with a vastly outnumbered force of 40,000 men. The fortified town of Segusio (Susa) blocked Constantine's way; Constantine stormed it but did not destroy or plunder it, thus encouraging other cities in northern Italy to surrender to him. Near Augusta Taurinorum (Turin) Constantine defeated an opposing force; Milan opened its gates to him. Constantine won again in another engagement which took place outside Verona; the rest of northern Italy came over to his side.

Maxentius had remained within the perceived security of Rome's Aurelian Walls. Perhaps Constantine would not be able to dislodge him, just as Severus and Galerius had failed earlier to dislodge Maximian. But, as Constantine's army approached, the people of Rome became restless and hostile, and Maxentius feared treachery. He consulted the Sibylline Books and learned that "on October 28 an enemy of the Romans will perish." October 28 was his *dies imperii*, he was superstitious, and he decided to offer battle outside the gates of the city. On a bridge of boats hastily constructed near the Milvian Bridge his army crossed over to the right bank of the Tiber. Here Maxentius suffered total defeat. His men were routed; thousands of them and Maxentius himself drowned in the Tiber. "An enemy of the Romans," from Constantine's perspective, had indeed perished.

The next day, 29 October, Constantine entered the city. Maxentius' body had been recovered, and Constantine had its severed head affixed to a pike and carried through the streets; later he sent it to Africa to deliver a forceful message there. The senate, which only a year ago had fawned upon Maxentius, now decreed *damnatio memoriae* for him while it elected Constantine senior Augustus.

Constantine was now the undisputed master of the West.

4

Constantine's conversion

When Diocletian and Maximian announced their retirement on 1 May 305 the persecution of the Christians was still in progress. The problem posed by the Christians, or rather by the emperors' insistence on religious conformity, had not been solved, and the members of the imperial college did not formulate or implement a uniform policy.

Soon after having been acclaimed emperor, Maxentius proclaimed toleration for the Christians in his territories; he did not, however, order the restitution of confiscated Christian property. A little later dissension among the Christians of Rome and concern for public order prompted him to intervene twice in the affairs of the church, banishing first Pope Marcellus and then Pope Eusebius. Constantinian propaganda, of course, would later denounce Maxentius as a "tyrant."

In April 311, on his deathbed, Galerius issued an edict in which Eusebius saw a "manifest visitation of divine providence" and which is sometimes called the "palinode of Galerius." In this remarkable recantation Galerius, while quite contemptuous of the Christians, acknowledged that the persecution had failed and grudgingly allowed the Christians once more to exist and to assemble in an orderly fashion; he did not provide for restitution. He bade the Christians pray for his well-being and for that of the state. Maximinus Daia, in his territories, ignored Galerius' edict and continued the persecution. Among his victims was the renowned presbyter and scholar Lucian, who was martyred on 12 January 312 at Antioch.

Constantine ended all persecution in his territories as soon as he came to power, providing not only for toleration but also for restitution. He acted formally and on his own, without consulting his imperial colleagues. He was not, however, at this time ready to embrace Christianity himself. His coinage evidences devotion first to Mars and then increasingly to Apollo, reverenced as *Sol Invictus*, "the unconquered sun." A panegyrist who addressed Constantine at Trier in 310, after the swift action against Maximian, will have us believe that Constantine on his way back from Massilia visited "the world's most beautiful temple," probably meaning the shrine of Apollo Grannus at Grand (near Neufchâtel, Vosges), and there experienced a vision: Apollo, accompanied by Victory, appeared to him and presented him with (four?) laurel crowns, promising a long and prosperous reign. The panegyrist's fulsome flattery continues: Constantine is the one whose coming "the divine songs of poets have prophesied." Significant is also that henceforth Constantine dissociates himself from the Herculians, the party of Maxentius; the new propaganda line claims that Constantine's father Constantius was a descendant of the emperor Claudius Gothicus (268–70), although the specific relationship was not made clear. Again the dynastic principle prevailed over the tetrarchic principle.

The next significant step in Constantine's religious development occurred in 312. Lactantius reports that during the night before the Battle of the Milvian Bridge Constantine was commanded in a dream to place the sign of Christ on the shields of his soldiers. The sign, it is widely believed, was the Chi-Rho, $\hat{P}$, although Lactantius' language on this point is not very clear. (Chi, X, and Rho, P, are the first two letters of the Greek form of "Christ;" in the monogram the two letters are written in ligature.) Constantine did as he had been told – his overwhelmingly pagan troops must have been puzzled – won the battle and from then on believed in the power of "the God of the Christians."

Twenty-five years later Eusebius, in his *Life of Constantine*, gives us a far different account, one which Constantine himself had given to Eusebius, and under oath: When Constantine and the army were on their march toward Rome – neither the time

22

nor the place is specified – they observed in broad daylight a strange phenomenon in the sky: a cross of light and the words "by this sign you will be victor" (*hoc signo victor eris* or τούτῶ νίκα). During the next night Christ appeared to Constantine and instructed him to place the heavenly sign on the battle standards of his army. The new battle standard became known as the *labarum*.

The two accounts are difficult to reconcile. Of the two, that by Lactantius is by far the more believable. Eusebius' account, while it fits the religious environment of his times, is suspect on account of both its timing and its substance. If the phenomenon was observed by the entire army, why then was it not more widely known? The phenomenon, if it did indeed appear, may have been a solar halo. Some scholars think that the *labarum* was not in use before 324.

In the Raphael stanze of the Vatican we may admire the great fresco of the Battle of the Milvian Bridge, as it captures for us one of the most significant events of all of ancient history. But, depicting a cross in the sky while the battle rages below, it is historically not correct. Such a picture is warranted neither by Lactantius nor by Eusebius.

We may be certain that in 312, before or during the battle, Constantine had an experience of some sort which was probably interpreted for him by Bishop Ossius (or Hosius) of Cordoba, who accompanied him, and which demonstrated to him the power of "the God of the Christians" and prompted him to profess Christianity. We may also be certain that this was more a matter of religious conviction than a matter of political expediency. Constantine was not a freethinker but a believer; he apparently had a deeply-felt need to place himself under the protection of a supreme deity. Professing Christianity gave him no advantage that could not be obtained by mere toleration, and the Christians were still very much a minority, especially in the West. Not too long after the capture of Rome, Constantine sent to the bishop of Carthage and to the proconsul of Africa letters which leave no doubt that he favored the Christian religion, subsidized the Christian church from public funds, exempted the clergy from public obligations, believed the proper worship of "the Deity" to be of vital importance for the welfare of the empire, and regarded himself as God's servant. But in what sense he became a Christian and how well he understood the Christian message is another question.

Late in 311 or early in 312, after the death of Galerius but before the Battle of the Milvian Bridge, Constantine betrothed his half-sister Constantia (one of the six children of Constantius and Theodora) to his fellow-emperor Licinius. Constantia was then eighteen years old at most, while Licinius was more than twice her age. This betrothal sealed a political understanding between the two emperors: Constantine was given a free hand against Maxentius, and Licinius a free hand against Maximinus Daia. Constantine, we have already seen, defeated Maxentius on 28 October 312 at the Milvian Bridge. Maximinus Daia suffered a defeat at Licinius' hands on 30 April 313 on the Campus Ergenus near Adrianople (in Thrace) and committed suicide at Tarsus (in Cilicia) a little later, perhaps in July. The Roman world was now in the hands of two masters, Constantine in the West and Licinius in the East.

In the meantime, probably in February 313, the two emperors had met in Milan. On this occasion the marriage of Constantia to Licinius took place. (It would be interesting to know by what rites this marriage was celebrated or how Constantia felt about this match.) At Milan, too, the two emperors agreed on a common religious policy. The agreement found expression, several months later, in the edict which is commonly but erroneously called the Edict of Milan. The text of this edict is recorded in Latin by Lactantius and in Greek by Eusebius. It is actually a letter addressed by Licinius to the governors of the provinces formerly controlled by Maximinus Daia. It instructs the recipients that all persecution of Christians is to cease, that confiscated Christian property, whether individual or corporate, is to be speedily restored and that all citizens, Christians specifically but also all others, are to be free to practise whatever religion they choose. It granted the religious toleration for which the Christian apologists had pleaded in the past. It did not alter the status of Christians in the West. It extended to Christians in the East the same protection which Christians in the West already enjoyed. It did not establish Christianity as a state religion; it did not commit Licinius personally to the Christian faith.

To return now to Constantine. There is no doubt that he was sympathetic to the Christians already before 312, that in 312 he committed himself personally to the Christian faith, and that in time his commitment to and understanding of that faith deepened. At the same time he seems to have realized that most of his subjects and especially the senatorial nobility in Rome were pagan, and he avoided offering offence to them. He still held the office of *pontifex maximus*. Christian symbols are slow to appear on his coins. The famous triumphal arch erected in his honor by the senate and completed in 315 is another case in point. There are no Christian symbols on the arch, and its language is religiously neutral: Constantine, it says, gained victory over the "tyrant" (Maxentius) "by the prompting of the Divinity (*instinctu divinitatis*) and by the greatness of his mind (*magnitudine mentis*)." He is the "liberator of the city" and the "establisher of peace."

It is now generally accepted that Constantine did not receive baptism until shortly before his death (see Chapter 11). It would be a mistake to interpret this as a lack of sincerity or commitment. In the fourth and fifth centuries Christians often delayed their baptism until late in life. This was true not only of Constantine but also of Constantius II, Theodosius I, and even St Ambrose. St Augustine comments on it in his *Confessions*. The practice was not, however, encouraged by the church.

A strange tale of Constantine's conversion is told by a hagiographic text of the early fifth century, the *Vita S Silvestri* or *Actus S Silvestri*. It knows nothing of Constantine's vision before the battle with Maxentius. Rather it turns Constantine, initially, into a pagan oppressor who orders all Christians to sacrifice. Pope Sylvester and the clergy withdraw to Mt. Soracte outside the city. Now Constantine suffers from leprosy, and pagan priests have advised him to seek a cure by bathing in the blood of infants, but his compassion does not allow him to follow their advice; he dismisses the infants, who have already been assembled to become his victims, and their mothers. He then, in a dream, is visited by the apostles Peter and Paul. He seeks out Pope Sylvester on Mt. Soracte, has the dream interpreted to him, and accepts Christianity. After a week spent in fasting and prayer he is baptized by Pope Sylvester in Rome and healed of his leprosy in the process. Sylvester then adds another exploit to his achievements: he defeats a fierce dragon which had been threatening the citizens of Rome. All this, supposedly, takes place right after Constantine has entered the city. Chronology alone bids us at once to dismiss this whole account as fiction: Sylvester did not become pope until 314.

But this strange tale enjoyed a long life and exerted considerable influence. It was accepted by the *Liber Pontificalis* (*Book of the Popes*), a collection of papal biographies first compiled *c.* 530. Later in the same century it underwent considerable embellishment, as happens so often in hagiography, in the *Chronicon* of Johannes Malalas: Constantine, it says, "was baptized by Sylvester, bishop of Rome, he himself and his mother Helena, and all his relatives and his friends and a whole host of other Romans." The *Chronicle* of Theophanes Confessor, dating from *c.* 810–15, claims to know that Sylvester baptized both Constantine and his son Crispus in Rome and dismisses reports to the contrary as a forgery of the Arians (on Arius and Arianism see Chapter 7); it seeks to prove its case by pointing to the "Baptistery of Constantine," meaning the Lateran Baptistery.

The *Breviarium Romanum*, published by Pope Pius V in 1568, and the *Martyrologium Romanum*, published in 1584 by Pope Gregory XIII, both assert that Constantine was baptized by Sylvester. In 1588 Pope Sixtus V caused to be erected in the piazza to the north of the Basilica of St John Lateran an Egyptian obelisk which had originally been brought to Rome and erected in the Circus Maximus by Constantius II but which had fallen long ago and lain neglected since then. On the base of the newly erected obelisk, the largest of Rome's obelisks, an inscription informs us that "Constantine was baptized here." In 1592 Cardinal Cesare Baronio, church historian and Vatican librarian, records in his *Annales Ecclesiastici* that Constantine was baptized by Sylvester. A last spirited defence of the thesis, flying in the face of all reason, was offered by a French author in 1906!

In the meantime the erroneous version of Constantine's baptism had inspired some outstanding works of ecclesiastical art. The small but beautiful Stavelot Triptych, dated *c.* 1165, in New York City's Pierpont Morgan Library, the thirteenth-century historical frescoes in the Oratory of St Sylvester at Rome's Church of I Santi Quattro Coronati and a stained glass window of the fifteenth century in the Church of St Michael in Ashton-under-Lyne (Greater Manchester) deserve mention,

26

among others. Most famous, however, is the "Baptism of Constantine" in the Raphael stanze of the Vatican.

Finally, the legend of St Sylvester made possible the famous eighth-century forgery which is known as the Donation of Constantine (more formally the *Constitutum Constantini*) and which was intended to boost the moral authority and secular power of the papacy. It is in the form of a letter purportedly written by Constantine in 315 (or 317?). Constantine acknowledges the primacy of the see of Rome over the patriarchates of Antioch, Jerusalem, Alexandria and Constantinople (but Constantinople had not yet been founded!). Grateful for having been baptized and for having been restored to health, he presents the Lateran palace to Sylvester and grants him and his successors dominion over Rome, Italy and "the West." He announces that he will build his own capital in the East (another anachronism!). He has given to Sylvester the right to wear various imperial insignia; he has ceremoniously held the reigns of his horse. One of the frescoes in the Oratory of St Sylvester at Rome's Church of I Santi Quattro Coronati, mentioned above, depicts Constantine doing just that. And Pepin, father of Charlemagne, is reported to have done the same for Pope Stephen II in 754. The Donation of Constantine was finally exposed as false in the sixteenth century.

We must turn our attention once more to the events of 312, to the decision which Constantine made in that year about religion. That decision, far from being a private matter, and its implementation within the remaining twenty-five years of Constantine's life, profoundly affected both church and state, religion and politics in the Roman world. Religion and politics were, of course, closely intertwined in the Roman world, just as they had been in the Greek world. Changes in one would inevitably bring changes in the other.

In the nearly three hundred years of its previous existence the church had periodically been subject to persecution; at the same time it had enjoyed independence. By his active involvement in its affairs, however benevolent, Constantine deprived the church of that independence. He deemed himself not only a divinely appointed ruler of the world but also a *koinos episkopos* (common bishop), that is, a general overseer and arbiter of church affairs. He used the church as an instrument of imperial policy and imposed upon it his imperial ideology. His desire for harmony and unity in the church took precedence over all other considerations. Clearly the church was now obliged to adopt a different attitude toward the empire, towards government authority, and toward military service. When the bishops assumed some judicial and administrative functions the church not only endorsed but became part of the apparatus of government.

The first Christians had been a humble and downtrodden minority. With the passing of time that gradually changed and Christians increasingly could be found among the urban middle class. By Diocletian's time we encounter them even at the court and elsewhere in imperial service, both civilian and military. But only under Constantine did the church acquire power and wealth. The original indifference towards wordly goods now gave way to the use of wordly goods in the service of the church. A decline in spirituality accompanied the process, and the rise of monasticism was, at least in part, a response to that decline and the increasing worldliness of the church.

The empire was affected not so much in its structure and institutions, which remained largely intact (see Chapter 10), as in its underlying ideology. The empire now became a copy of the Kingdom of Heaven, and the emperor God's viceregent on earth. Constantine had become the founder of a Christian empire. Edward Gibbon attributed the decline and fall of the Roman empire to four causes; one of these he called "the triumph of barbarism and religion," meaning the triumph of the church. The juxtaposition of the two terms says much about Gibbon's prejudice, a prejudice not shared by many today.

5

Constantine as the sole ruler of the West

The Roman government had always regarded the oversight and regulation of religious affairs as one of its legitimate functions, an effort to maintain the *pax deorum*, the harmonious relationship between the Roman people and the gods. The emperors themselves, beginning with Augustus, had held the post of *pontifex maximus* and thus had stood at the head of the religious establishment. Constantine, therefore, was fully within the Roman tradition when he announced his support of Christianity, paid subsidies to the Christian church and granted immunities to its clergy. Little did he know that his actions were to mark the beginning of one of the great problems of post-classical Western Civilization, the relationship of secular authority to spiritual authority, of state to church. And if he thought that the church might be of help in unifying the empire he was soon disappointed. He had not anticipated that disputes within the church would absorb a major portion of his time and energies.

In April of 313 Constantine received a petition addressed to him by certain Christians in North Africa and forwarded to him by the proconsul Anullinus. The petitioners had a grievance against Caecilian, the bishop of Carthage, and asked Constantine to appoint from among the bishops of Gaul judges to hear their case (from Gaul, because there had been no persecution there, they said). It is to be noted that the petitioners addressed themselves not to the bishop of Rome, but to the emperor, thus recognizing him as arbiter of church affairs. It was not, however, the first time that an emperor had acted as a judge in church affairs: forty years earlier Aurelian had been asked to judge a dispute between the church of Antioch and Paul of Samosata, its deposed bishop, over church property.

But who were these petitioners and what was their grievance? To answer this question we must revisit the Great Persecution of Diocletian. Not all the bishops and clergy had responded to that persecution with the same degree of faith and fortitude. Some had openly defied the authorities, sometimes to the point of inviting martyrdom; others had forsaken their faith and had meekly surrendered the scriptures to be burned, thus becoming *traditores* (from the Latin *trado*, to surrender); yet others had followed a middle course by going into hiding or by surrendering heretical books rather than the scriptures. When the persecution had run its course there arose a conflict between rigorists, who advocated strict discipline in dealing with those who had been lacking in faith, and moderates, who took a more forgiving attitude. In North Africa this battle was fought between Secundus, the primate of Numidia, and Mensurius, the metropolitan of Carthage, representing the rigorist and moderate party respectively. Nor was this simply a theological dispute or a disagreement among the leaders of the church. Rather it took on aspects of a class conflict, since supporters of the rigorist party were to be found primarily in the rural population and in the lower urban class, while the moderates drew strength primarily from the urban middle and upper class.

When Mensurius died in 311, his archdeacon Caecilian, also of the moderate party, quickly prevailed upon three bishops – the minimum number to make an election valid under church law – to elect him bishop and upon one of them, Felix of Aptunga, to ordain him. The rigorists were by no means willing to accept this fait accompli. Also the flames of the conflict were fueled by a certain Lucilla, a wealthy Spanish lady, whose enmity Caecilian had incurred by censuring her for ostentatiously venerating the relic of a martyr. And thus in 312 Bishop Secundus held a council of seventy bishops – the number seventy has a rich tradition in Jewish and Christian practice – in Carthage; this council ruled that Caecilian had been elected uncanonically, deposed him, and elected a certain Majorinus, the lady Lucilla's chaplain, in his stead. But Caecilian could not be easily dislodged, and it is he who was benefiting from Constantine's largesse, as mentioned in the preceding chapter.

In formulating his response to the petition Constantine probably had the assistance of Bishop Ossius (or Hosius) of Cordoba, who had joined the court as an advisor in 312. Constantine chose three Gallic bishops to serve as judges in the dispute; these were Reticius of Autun, Maternus of Cologne, and Marinus of Arles. He also asked Bishop Miltiades of Rome to preside. Caecilian was invited to defend himself in person and to bring ten of his supporters with him; his opponents also were invited to send ten representatives. Miltiades, in turn, added fifteen Italian bishops to the panel of judges, making it in fact a synod of bishops, the Synod of Rome. In the meantime Majorinus had died; the rigorists replaced him by a certain Donatus and thus became known as Donatists. The synod convened on 2 October 313, in Rome's Lateran Palace, which Constantine had given to the bishop of Rome as his residence. It took the council only three days to find in favor of Caecilian. The dissidents did not give up; they appealed, again not to Bishop Miltiades, but to the emperor. The ordination of Caecilian was invalid, they claimed, because Felix, the bishop who had consecrated him, had been a *traditor*, that is, one who had surrendered the scriptures. They even impugned Bishop Miltiades because he had been deacon, ten years earlier, to Marcellinus, the lapsed (briefly) bishop of Rome.

Constantine responded to the appeal by convening a larger council of bishops which was to meet on 1 August 314, at Arles in southern Gaul under the presidency of Marinus, the bishop of the host city. To the participants he generously offered the services of the imperial transport service. Thirty-three bishops, mostly from Gaul, but also including five from Britain, and thirteen other clergy assembled. Among them was Agritius of Trier, who was mentioned in Chapter 3 above, but not Miltiades. Before the council completed its business Miltiades had passed away and had been replaced by Sylvester, whom tradition erroneously credits with the conversion and baptism of Constantine, as we have seen in Chapter 4 above. In his *Life of Constantine*, in the context of events transpiring in the years 312–15, Eusebius reports approvingly that Constantine convened synods of God's servants, sat with them in their

31

assembly and participated in their deliberations. This certainly suggests, but does not fully prove, that Constantine himself attended the council. The council must have been in session for a number of weeks, since it considered not only the Donatist problem, but other matters as well, such as celibacy of the clergy, consecration of bishops, the date of Easter, the thorny issue of rebaptism, and Christians serving in the Roman army. In the end the emperor dismissed the assembled bishops, calling them his *fratres carissimi* (dearest brethren), but reportedly tiring of them.

At the conclusion of the council the bishops reported on their labours and their decisions to Pope Sylvester; part of their letter and the twenty-two canons are extant. They had found, as had their colleagues meeting at Rome in the preceding year, against Donatus and in favour of Caecilian. The Donatists, however, did not submit. Constantine issued another imperial ruling against them in the autumn of 316. His attempts to suppress them show his readiness to use the powers of the state in an effort to end a dangerous schism in the church, but did not succeed. Although the so-called Edict of Milan had proclaimed religious toleration throughout the empire, he ordered that their property be confiscated and their leaders be exiled. In 321 he decided to abandon the use of force against them and turned his attention to other matters. A separatist Donatist church possessed considerable strength in North Africa through the fourth and into the fifth century. In 336 no fewer than 270 Donatist bishops assembled for a council. St Augustine's theology was shaped in part in response to the Donatist challenge. Only the Muslim invasion in the seventh century finally wiped out the last traces of Donatism.

From Arles the emperor departed for the Rhine frontier to undertake an autumn campaign against the German tribes. Eusebius, with his usual hyperbole, reports that Constantine won victories over "all barbarian nations," and some of the coins minted in 314–15 celebrate him as *victor omnium gentium*. On 29 October, 8 November and 30 December 314 he can be shown from subscriptions in the *Codex Theodosianus* (the codification of Roman Law undertaken by the emperor Theodosius II in 438) to have resided in Trier. He was in Trier also when he embarked on the consulship for the year 315, as we learn from a gold coin issued to celebrate this event. During an

32

earlier campaign he had built a bridge across the Rhine, as we have seen in Chapter 3 above. In 315 he completed and dedicated a fort at Divitia (Deutz) on the right bank of the Rhine, opposite Cologne, as we learn from the extant dedicatory inscription.

This must have been before he left for Rome, there to observe the beginning of his *decennalia* (the tenth anniversary of his acclamation). He arrived on 18 or 21 July. The famous Arch of Constantine, referred to in Chapter 3 above, was completed in time and dedicated as part of the festivities. There were circus races, games and donatives, but Constantine pointedly omitted the traditional sacrifices to the pagan gods. To high-ranking imperial officials he presented special silver *decennalia* medallions with Christian symbols on them. On 27 September he left Rome to return to Gaul. On 19 October 315 he can be shown to have been in Milan. The winter of 315/16 he spent in Trier, rejoining his sons Crispus and Constantine II, his wife Fausta, and his mother Helena.

It will be convenient next to examine Constantine's building programme in the city of Rome and to place it in its political and religious context. During the reign of Diocletian the city of Rome had lost much of its political significance and ceased to be an effective centre of power. The new imperial residences, such as Nicomedia or Trier, were now the real centres of power. While Rome was still the seat of the senate, that body was now limited largely to honorific and ceremonial functions; even the urban prefect was appointed by the emperor. On the other hand the senators represented wealth and venerable tradition; many of them served as consuls, prefects, or provincial governors. Their presence made Rome the only legitimate capital, which was still hallowed by its long history and celebrated by the poets. Diocletian had acknowledged this by choosing to observe there his *vicennalia*, as already mentioned. Constantine followed suit by celebrating both his *decennalia* and his *vicennalia* in the city, but after 326 he did not visit the city again.

The building activity of the emperors also tells us that the symbolic significance of the city was by no means lost upon them. Diocletian restored the *curia* (senate hall) and the Basilica Julia in the Roman Forum, both of which had been destroyed by fire, and built the huge Baths which bear his name and accommodate today within their walls the National Museum

(Museo Nazionale Romano delle Terme) and the Church of S Maria degli Angeli (St Mary of the Angels). Maxentius left to the city of Rome a remodeled Temple of Venus and Roma, the rotunda which is erroneously known as the Temple of Romulus (after his son Romulus, not *the* Romulus), the unfinished Basilica Nova (also known as the Basilica of Constantine), all these in the Forum, and on the Appian Way a circus and a mausoleum. He also raised the Aurelian Wall to nearly twice its original height; how ironic that it did not save him on that fatal day in October 312.

Constantine could not do otherwise, although he does not seem to have been fond of the city. On the Quirinal Hill, where the presidential palace stands today, he built a bath; from it the statues of the two Dioscuri (Castor and Pollux), flanking an obelisk, survive. In the Roman Forum he completed the vast Basilica Nova which Maxentius had left unfinished, adding a second apse on the north side and a second entrance on the south side. This building was the last of the great secular basilicas in Rome and has rightly been called a masterpiece of engineering and architecture. It consisted of a central nave and two side aisles; only the north aisle stands today. The roof of the nave was cross-vaulted and supported by eight monolithic columns; one of these has survived, standing today in the Piazza S Maria Maggiore. To his mother Constantine assigned the Sessorian Palace, impressive remains of which still stand today and one large hall of which was later (under Constantine's sons, this writer believes) converted into the Church of S Croce in Gerusalemme. His mother is known to have restored the baths which were then called after her but no longer stand. The massive four-sided arch known as the Janus Quadrifons, in the Forum Boarium, probably also dates from Constantine's reign.

Constantine's church foundations interest us especially (Figure 3). The four "patriarchal churches" of Rome are S Giovanni in Laterano (St John's at the Lateran), S Pietro in Vaticano (St Peter's in the Vatican, to distinguish it from S Pietro in Vincoli), S Paolo fuori le mura (St Paul's outside the walls), and S Maria Maggiore (St Mary Major). The first two of these were built on Constantine's orders and with his support, while the third is less securely associated with him and the fourth postdates him by a century.

34

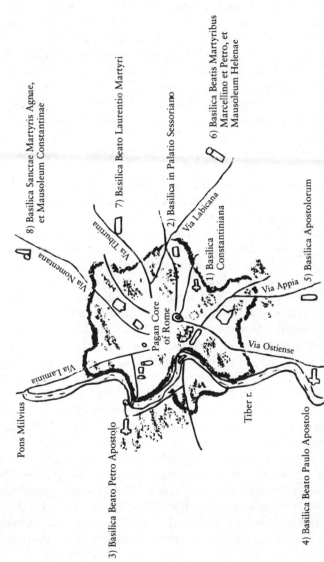

8) Basilica Sanctae Martyris Agnae,
et Mausoleum Constantinae

7) Basilica Beato Laurentio Martyri

2) Basilica in Palatio Sessoriano

6) Basilica Beatis Martyribus
Marcellino et Petro, et
Mausoleum Helenae

Via Tiburtina

Via Nomentana

Via Labicana

1) Basilica
Constantiniana

Pagan Core
of Rome

Via Appia

5) Basilica Apostolorum

Via Laminia

Via Ostiense

Tiber r.

Pons Milvius

3) Basilica Beato Petro Apostolo

4) Basilica Beato Paulo Apostolo

Figure 3 Map of the Christian basilicas of Constantinian Rome. From Charles M. Odahl, *Early Christian Latin Literature* (Chicago, 1993), p. 139.

Courtesy of Charles M. Odahl and Ares Publishers, Inc.

Among the imperial properties which fell into Constantine's hands when he became master of Rome was the Lateran Palace, so named after the family which at one time had owned it but had lost it in the days of Nero, the Laterani; it was also known as *domus Faustae*, because it had been part of Fausta's dowry. This palace Constantine gave to Pope Miltiades as a residence, and here was held the synod which dealt with the Donatist problem. Next to the palace stood the barracks of the *equites singulari*, a military unit which Constantine disbanded. These barracks Constantine ordered torn down to make room for his first Christian foundation, which was originally called the Basilica Salvatoris or the Basilica Constantiniana but later the Basilica of S Giovanni in Laterano and work on which was begun before 315, perhaps as early as the autumn of 312. It was a large building, measuring 98 × 56 m., and featured a nave, double side aisles, an apse, and a wooden roof. Little of the original construction survives. Next to the basilica he built the large octagonal baptistery which became a model for a number of later baptisteries. S Giovanni in Laterano was and still is the great mother church of all Christianity, the cathedral church of Rome, outranking St Peter's.

Work on "Old St Peter's" (to distinguish it from the present structure) was begun between 315 and and 319 and completed around 329. This basilica was even larger than that of S Giovanni, being 120 m. long. It also was different in plan. It, too, had a central nave and double side aisles, but unlike S Giovanni it featured an enclosed atrium (a feature also of several other fourth-century churches), a narthex and a transept. A dedicatory inscription in gold lettering was placed over the triumphal arch. The episcopal throne and seats for the clergy were in the apse. The interior colonnades featured 96 columns. Four columns from Old St Peter's were reused in 1610–12 by the architects Giovanni Fontana and Flaminio Ponzio in the construction of the elaborate Fontana Paola on the Janiculum Hill. In the centre of the transept, that is, in the crossing, was the *martyrium*, the shrine of St Peter, for this basilica was built deliberately over the apostle's tomb (and in the process an ancient cemetery was covered over). The *Liber Pontificalis* (*Book of the Popes*) reports that Constantine placed a huge cross of pure gold above the tomb of St Peter; it also records, somewhat imperfectly, the inscription on this cross, and according to this inscription Constantine and Helena were the joint donors. The existence of such a cross on the tomb of St Peter is confirmed by the carvings on the rear side of a fifth-century ivory casket found at Samagher near Pola in Istria and now to be seen in the Archaeological Museum of Venice.

On the Appian Way, over the Catacomb of St Sebastian, Constantine erected the Basilica Apostolorum, so called because of the belief that at one time, during the persecutions, the relics of the Apostles Peter and Paul had been taken there for safety's sake. The ancient church, U-shaped in plan, with an ambulatory around the apse, was considerably larger than the present church of S Sebastiano. Another funerary basilica of Constantinian date and of similar design was that of St Lawrence on the Via Tiburtina.

On the ancient Via Labicana, today's Via Casilina, about three miles from the Porta Maggiore, the ancient Porta Labicana, Constantine erected a basilica in honor of the martyrs Marcellinus and Peter. Attached to this basilica was the vaulted rotunda which Constantine at one time had intended as a mausoleum for himself and his family. The basilica no longer stands and has been replaced by a church of more recent times. But of the mausoleum there are substantial remains, known as Tor Pignattara, from the clay pots ("pignatte") which are built into its fabric. A rectangular niche in the mausoleum wall, facing the entrance, once accommodated the porphyry sarcophagus of Helena which is now in the Sala a Croce Grece of the Vatican Museum.

It is often pointed out that Constantine's Christian foundations were all located on imperial property or outside the city walls, far removed from the centre of the city with its pagan temples. All his foundations were richly endowed and no expense was spared, but the churches were more resplendent on the inside than on the outside. The ambiguity of his building programme thus reflected the ambiguity of his political programme: he desired to promote Christianity; he also desired to avoid an open conflict with the pagan party. For the history of architecture it is significant that Constantine chose for his churches a style of building that had long served secular purposes, the basilica.

6

The conflict with Licinius

Early in 313 Constantine and Licinius met at Milan, as we have seen in Chapter 4, and agreed on a common religious policy. They also, while not dividing the empire, agreed on their respective spheres of control. Licinius did not press a claim on Italy, which was rightfully his by virtue of his original appointment but which Constantine had seized from Maxentius. Licinius went on to eliminate Maximinus Daia and to campaign first on the Persian frontier and then against the Goths. Constantine went on to campaign on the Rhine frontier and to celebrate his *decennalia*. About July of 315 Constantia bore to Licinius a son who was named Valerius Licinianus Licinius; the birth of this child should have served further to strengthen the ties between the two emperors. In 312, 313 and 315 Constantine and Licinius jointly held the consulship, each for the second, third and fourth time respectively; in 314 they allowed two senators to serve as consuls. The Augusti appear together on some of the coins minted in these years. The senate, after the Battle of the Milvian Bridge, had designated Constantine as *Maximus Augustus* (senior Augustus), and he is thus referred to in some inscriptions. But the decorative programme of the Arch of Constantine suggests harmony and equality between the rulers, for they are there depicted no fewer than eight times in strict parallelism. Nevertheless the relationship between the two men was a strained one. Deep suspicions must have existed on

38

both sides, for both men had demonstrated by their past conduct that their ultimate goal was sole power. The agreement of 313 had been born of necessity, not of mutual good will. It was inevitable that hostilities eventually should erupt.

This was in 316, not in 314 as one of our sources erroneously reports and as some scholars erroneously assert. Constantine's movements recorded for that year yield a clear picture: in January he can be shown to have been in Trier, in March in Châlons-sur-Saône, in May in Vienne, in August in Arles, and in September in Verona, while Licinius appears to have been in the Balkans. At about the same time the image of Licinius disappeared from the bronze coinage of the mints of Arles, Trier, Rome, and Ticinum, suggesting a break in relations. On 8 October the two armies met at Cibalae in Pannonia; hence this war became known as the *bellum Cibalense*. Licinius' army, in spite of numerical superiority, suffered heavy casualties, and Licinius fled from the field of battle, first to the relative safety of Sirmium, and then to Adrianople. It is reported specifically that he conveyed his wife, his son and his treasury to safety. A second battle ensued, perhaps in January of 317, on the *campus Ardiensis* in Thrace; neither side could claim a clear victory in this second engagement.

Under the terms of a settlement Licinius ceded to Constantine all of his European provinces except Thrace but retained his position as Augustus. On 1 March 317, at Serdica (modern Sofia), Constantine announced the appointment of three Caesars. These were his son Crispus (by Minervina), about twelve years old, his son Constantine (by Fausta), born on 7 August 316, and Licinius' son, the younger Licinius, born to him by Constantia and twenty months old. While these appointments clearly gave an advantage to Constantine, they were meant to heal the break between the two emperors. *Concordia Augustorum*, proclaimed some of the coins on the occasion.

The consulship was again held by two senators in 317. In 318 the consulship was shared by the Augustus Licinius, for the fifth time, and by the Caesar Crispus, for the first time. In the following year, by reciprocal arrangement, the two consuls were the Augustus Constantine, for the fifth time, and the Caesar Licinius, for the first time. In 320 Constantine abandoned the principle of reciprocity and of shared honors when he held the consulship for the sixth time and took his young son Constantine as a colleague. For 321 Constantine again claimed both consulships for his own family, appointing his sons Crispus and Constantine, the Caesars, both now for the second time. Not surprisingly these two were not recognized in the East, where the two Licinii claimed the consulship for themselves. For both 322 and 323 Constantine appointed distinguished senators, and for 324 once again his sons Crispus and Constantine; none of these were recognized by Licinius. The *concordia Augustorum* was breaking down.

In matters of religion, too, there was a parting of the ways. Constantine did not, after the Battle of the Milvian Bridge, immediately or entirely abandon his allegiance to the sun god. The reverse legend *Soli Invicto Comiti*, "to the Unconquered Sun his Companion," appears on his coins as late as 320. When, in 321, he ordered the law courts not to conduct business on Sunday he referred to Sunday not as "the Lord's Day," but as the "the day celebrated by the worship of the Sun." Nevertheless he regarded himself as the servant of the Christian God, who "had committed to his [Constantine's] care the government of all earthly things." There was a gradual deepening of his commitment to and faith in the Christian gospel. In 315 the mint of Ticinum produced a silver medallion depicting Constantine with the Christian monogram on his helmet (Figure 4). The coinage is thus, like Constantine's policy, ambiguous, giving evidence not of a sudden conversion but only of a gradually changing attitude.

We have other evidence to consider. Bishop Ossius was counted among Constantine's most trusted advisors. To another Christian, namely the learned Lactantius, Constantine entrusted the education of his son Crispus. (Unfortunately we know neither exactly when Lactantius assumed his duties nor what Crispus' position was on matters of religion.) Writing in his *On the Deaths of the Persecutors*, c. 314, Lactantius calls Constantine "the first of the Roman emperors who, having repudiated errors, acknowledged and honored the majesty of the unique God." Constantine's *Oration to the Assembly of the Saints*, delivered on a Good Friday between 317 and 324, or, it has been suggested, in 325 at Antioch, also tells of his continuing religious development, betraying the influence of Lactantius. And then there is his legislation, clearly favouring the church: in 318 he extended the judicial authority of the bishops;

Figure 4 Silver medallion of Constantine, minted at Ticinum in 315.

From *Reallexikon für Antike und Christentum* V (Stuttgart: Anton Hierseman Verlag, 1957), p. 327.

Courtesy of Franz Joseph Dölger-Institut, Bonn.

in 320 he removed the disabilities which had been imposed long ago by Augustus on celibates; in 321 he legalized bequests to the church; in the same year he gave permission for the manumission of slaves to take place in church.

Licinius followed a different path. He had agreed to a policy of toleration and abided by that policy for a time. His wife was a devout Christian, as we know from correspondence between her and Eusebius of Caesarea. But he himself was never converted. In 313, before the battle against Maximinus Daia, he had his soldiers recite a monotheistic, but not Christian, prayer. His coinage evidences devotion to Jupiter Conservator, in the tetrarchic tradition founded by Diocletian. He suspected his Christian subjects of being supporters of Constantine. In time he had recourse to various oppressive measures: he dismissed Christians from the imperial service, ordered the Christians to meet for worship not in their churches, but in the open, outside the city gates, and in separate assemblies for men and women, and forbade the bishops to visit one another's cities, thus making it impossible for them to hold councils or to ordain other bishops. He did so although Eusebius, the bishop of Nicomedia, was a favorite of Constantia and had considerable influence at the court.

There appear to have been some cases of arrest and execution, too. The much-celebrated martyrdom of the Forty Martyrs of Sebaste – Christian soldiers who froze to death rather than sacrificing to the pagan gods – is said to have occurred under Licinius. Licinius' stance allowed Constantine to cast himself in the role of liberator of the oppressed Christians in the eastern half of the empire and to present the forthcoming war as a crusade in the cause of the Christian God. Licinius, according to Eusebius' *Life of Constantine* (not an unbiased witness), was "confident in the aid of a multitude of gods" to the bitter end.

While tension between the two emperors grew, Constantine gathered his forces in the Balkans, residing much of the time at Sirmium or at Serdica. In 323, in the course of campaigning against the Goths (or Sarmatians?), who had crossed the Danube and invaded Roman territory, he violated Licinius' territory and thus created a *casus belli*. Hostilities commenced in 324; both sides had amassed armies well in excess of 100,000 men. Constantine defeated Licinius twice: on 3 July at Adrianople, and on 18 September at Chrysopolis; victory, so it was now claimed, had come to his army by the miraculous power of the *labarum*. Licinius survived both battles, withdrew to Nicomedia and surrendered to Constantine shortly thereafter. Constantine entered Nicomedia in triumph; he had now become the sole and undisputed master of the Roman world.

A substantial contribution to Constantine's victory had been made by his son Crispus, who had already proven himself in military operations against the Franks and Alamanni. This time, in spite of his youth, he was entrusted with the command of a fleet assembled at Thessalonike and consisting of 200 warships and 2000 transports (the latter needed to ferry the troops across

42

the Bosporus). He defeated Licinius' less than brilliant admiral Abantus (or Amandus) at the entrance of the Hellespont, secured the Hellespont, the Propontis, and the Bosporus, and blockaded Byzantium, which fell after a siege of two-and-a-half months.

Constantia appears to have stood by her husband all these years, which must have been difficult, given the differences betwen husband and wife in age and religious outlook. After the Battle of Chrysopolis she betook herself to Constantine's headquarters and pleaded for her husband's life. Constantine, yielding to his sister's pleas, promised to spare Licinius' life. Licinius then placed himself in the victor's hands, and Constantine ordered him to live as a private citizen at Thessalonike. Only a few months later, however, in the spring of 325, he ordered Licinius to be put to death. Three of our ancient sources tell us that in so doing he broke a solemn oath. One of these sources, the historian Zosimus, is consistently hostile to Constantine, but the same is not true of the other two, Eutropius and St Jerome. Eusebius, always endeavouring to put Constantine's deeds in a favourable light and without offering any details, simply says that Constantine subjected Licinius to "the just punishment of death." In the next century the historian Socrates claims to know that Licinius had plotted with some barbarians to renew the war – not a very likely story at all.

Even the younger Licinius, a child about ten years of age, fell victim to Constantine's anger or suspicions. We have three reports of his death; taken together they strongly suggest that it occurred either in 325 or in 326. One wonders how Constantine justified the killing of an innocent child, his own nephew, and how Constantia could have failed to rebuke him bitterly. She survived the loss of husband and son by a number of years, occupying a position of honor and influence at Constantine's court and holding the rank of *nobilissima femina*. She even attended the Council of Nicaea and was a friend of the Arian cause. (On these matters see the following chapter.) When she died, perhaps in 330, Constantine was at her side. After her death she was even honored by a commemorative coin and by having a town named Constantia named after her (Maiuma, the port of Gaza in Palestine).

There was still another victim of Constantine's vengeance, twelve years later. This was a second, illegitimate son of Licinius, who was born to him by a slave woman and is not to be identified with the son born to him by Constantia, although not a few scholars, beginning with Otto Seeck, have done just that. This son, although not identified by name, is the subject of two of Constantine's rescripts (laws). The first of these was received in Carthage on 29 April 336 and ordered that "the son of Licinianus," who had "ascended to the summit of dignity" (probably to the rank of *vir nobilissimus*), was to be stripped of all honors and property, to be scourged and bound, and to be returned to the status of his birth. From the second rescript, received in Carthage on 21 July 336, we learn that the unfortunate "son of Licinianus" had escaped but had been apprehended; he was now to be bound and to be consigned to work in the *gynaeceum* (imperial weaving establishment) in Carthage.

Having secured his newly acquired eastern provinces, Constantine lost no time in ordering religious affairs. A letter which he circulated to the provincials, still in the autumn of 324 went significantly beyond the so-called Edict of Milan. No longer was there a pretence of religious neutrality; no longer did he disguise his religious sympathies; rather he stressed the truth of Christianity and the errors of paganism. God's power had granted him victory, so he wrote, and put an end to impious tyranny. He ordered the release of all Christians condemned to forced labor, the return of all Christian exiles to their homes, and the restoration of all property, private or corporate, confiscated from the Christians. He encouraged the bishops to repair damaged churches and to build new ones where needed. He went further yet when he prohibited sacrifice to the pagan gods, consultation of the oracles and the dedication of new cult statues. Enforcing these prohibitions was, of course, another thing; paganism did not die a sudden death. But Christianity was now not only the favoured but the established religion of the empire.

7

The Arian controversy, the Council of Nicaea and its aftermath

During its first three hundred years the Christian church not only had to endure periodic persecutions; it was also threatened from within by various heretic or schismatic movements such as Docetism, Gnosticism in its various manifestations, Montanism, Sabellianism and Donatism. In the face of such challenges there was a need for the church to tighten its organization, to establish its canon of scripture and to define more clearly its doctrine. One point which needed clarification was the church's teaching on the Godhead. In the West, never much given to philosophical or theological speculation, the church was largely in agreement on this subject, believing in the unity of the Godhead and in the equality of the three persons within it. Not so in the East, where Origen especially had given to the study of theology a more philosophical, specifically Platonic, character. The question that had been raised early in the fourth century was that of the exact relationship of the Son to the Father. As the distinguished German theologian Adolf von Harnack expressed it in 1905, "Is the divine that has appeared on earth and reunited man with God identical with the supreme divine, which rules heaven and earth, or is it a demigod?" The dogma of the Trinity, the climax of the doctrinal development of the early church, was eventually to answer that question.

The church of the first and second centuries had answered Harnack's question in the former rather than in the latter sense.

"Christ . . . who is over all, God blessed forever," we can read in Romans 9:5 (King James Version; different punctuation can give a different meaning to this passage). The second-century homily known as 2 Clement begins thus, "Brethren, we ought to think of Christ as of God, as the judge of the living and the dead." And in the *Martyrdom of Polycarp* (155 or 156) we find this: "For him (Christ), as the Son of God, we adore; the martyrs, as disciples and imitators of the Lord, we reverence." But there was also the "adoptionist" view, which held that divinity was conferred on the man Jesus at a specific time, such as his baptism or resurrection; Acts 2:32–36 can be interpreted in this sense. And the so-called Monarchians (this term includes the Sabellians) of the second and third centuries held that Father, Son, and Holy Spirit were only different "modes" rather than separate identities.

It is not by coincidence that the issue came to a head in Alexandria, some time after 312. Bishop Alexander invited several presbyters to offer their interpretation of a biblical passage, apparently Proverbs 8:22–31, "The Lord created me at the beginning of his work," etc. The learned (but somewhat arrogant, it is said) Arius, a former student of Lucian of Antioch, offered his doctrine of Christ as a creature: Christ, before he was begotten or created, did not exist; there was a time when he was not. Christ thus is not co-eternal with the Father. Father, Son, and Holy Spirit are three distinct *hypostaseis* (Greek *hypostasis* or *ousia*; Latin *substantia*; English "substance" or "essence"). The Son was subordinate to the Father, an intermediary, ranking first among creatures.

In about 318 Bishop Alexander assembled in council almost one hundred bishops from Libya and Egypt, who condemned Arius' doctrine and excommunicated him. Arius appealed for help both to Eusebius of Caesarea and to Eusebius of Nicomedia; both bishops interceded on his behalf, much to the annoyance of Alexander. Arius first proceeded to Caesarea and then found refuge at Nicomedia, where not only the bishop, a fellow-"Lucianist," but also the empress were kindly disposed toward him (319); the dispute continued.

When Constantine, before the end of 324, became aware of the dissension, he dispatched his trusted spiritual advisor Ossius to Alexandria in a quest for unity. Ossius carried with him a letter which Constantine had addressed both to Arius (who had

46

returned to Alexandria) and to Alexander. In this letter Constantine expressed his dismay and bitter disappointment at what he had learned. He had hoped, he wrote, to find through them a remedy for the errors of others (meaning the pagans), but these hopes had now been dashed. The cause of their dispute he found trifling and insignificant; such questions should never have been posed, nor were they worthy of a response. "Give me back peaceful nights and days without care," he pleaded.

Ossius' mission failed. On his way back to Nicomedia he stopped in Antioch, early in 325, there to join a group of more than fifty bishops assembled to choose a new bishop for the city; their choice had fallen on Eustathius, a staunch anti-Arian. Ossius had no difficulty in getting the assembled bishops to condemn Arius and his doctrine; they anathematized "those who say that the Son of God is a creature" or "that there was (a time) when he did not exist." Three dissenting bishops, including Eusebius of Caesarea, were excommunicated but were given an opportunity to rehabilitate themselves at a future council.

Constantine now responded to the crisis by summoning what has become known as the First Ecumenical Council of the church (Figure 5), designating first Ancyra and then Nicaea as the meeting place. Between two hundred and three hundred bishops responded to the call; the exact number can no longer be determined, and 318, sometimes given as the correct number, is certainly fictitious. Those who came were mostly from the East. Pope Sylvester, pleading ill health, sent two deacons to represent him. Italy, Gaul and Africa each sent one bishop only; Britain and Spain sent none, although Ossius attended in his capacity as Constantine's advisor (and took precedence even over the Pope's representatives). Alexander of Alexandria was there, of course; so was Eustathius of Antioch and Marcellus of Ancyra, another staunch opponent of Arius. Some of the assembled bishops were confessors who had suffered under the recent Great Persecution, such as Paphnutius of Egypt and perhaps Spyridon of Cyprus. Also in attendance was Athanasius, Bishop Alexander's protégé, deacon and secretary, whose career soon would dominate the history of the controversy until his death on 2 May 373.

The opening session was held on 20 May 325 in the great hall of the palace at Nicaea. Constantine himself was there, resplendent in his imperial robes, and gave the opening speech. He

Figure 5 Icon of the First Ecumenical Council. Artist unknown, seventeenth century.
Courtesy of the Museum of Zakynthos, Greece.

urged harmony and deplored dissension, which he deemed worse than war or disaster. And he continued to not only attend but also to participate in the deliberations, which can hardly have been conducive to open and free discussion. Early in the proceedings Eusebius of Caesarea, still under the cloud of his excommunication at Antioch, introduced the creed of his church at Caesarea (see Appendix IV). It found acceptance only with some highly significant additions. Constantine himself, probably on the advice of Ossius, proposed the addition of the word *homoousios* (Latin *consubstantialis*; English "of the same essence" or "of the same substance"). Anathemas were added against any who said that "there was (a time) when he (Christ) was not," or that "before being born he was not," or that he was created, or that he was of another substance or essence than the Father. The creed that emerged (see Appendix IV) was thus substantially different from the one which Eusebius had introduced.

All the bishops but two signed. An embarrassed Eusebius sent a letter to his church at Caesarea, putting the events and his own actions in as good a light as he could. Arius refused to yield and was exiled, as were the two bishops who had refused to sign the creed. Eusebius of Nicomedia and Theognis of Nicaea signed, but protested against the excommunication of the two dissident bishops; three months later they were themselves deposed and exiled, as we shall see.

The *acta* (minutes) of the deliberations are not extant, but the twenty canons of the council are. Among the issues considered by the council was the date of Easter, a matter which had also occupied the bishops at Arles in 314. All churches, it was decided, henceforth should follow the calculations already employed in most parts of the empire, but not in Syria, Palestine and Illyricum. Constantine himself communicated this decision to the Christian churches, voicing some violently anti-Semitic sentiments as he did so. The council also attempted to heal the Melitian schism, which had divided the church of Egypt. Bishops and clerics were again restrained from moving from one city to another (canon 15); they were also prohibited from lending money at interest (canon 17). On his *dies imperii*, 25 July, the emperor invited the bishops to a banquet in the palace. The next day he gave a farewell speech, once again commending concord and cooperation.

The Nicene Creed, although subsequently revised at the Council of Constantinople in 381–2 (see Appendix IV), became the touchstone of orthodoxy. At its core is the *homoousion*, that is, the doctrine of consubstantiality. In the Latin-speaking West the Creed became known as the *symbolum Athanasianum* or the *Nicaenum*.

Constantine took pride in the Council of Nicaea and the work which it had accomplished under his leadership. But there was more work to be done for unity within the church. Shortly after the council he issued an edict against various minor heretic or

49

schismatic groups such as the Valentinians, Marcionists (both Gnostic sects), Novatians (followers of the rigorist Novatian), Cataphrygians (the Montanists of Phrygia) and Paulianists (followers of Paul of Samosata). But Constantine's real problem continued to be with the Arians. In Alexandria there were several priests who clung to the condemned doctrine of Arius. Constantine had them deported to Nicomedia. There they were supported in their stance by Eusebius of Nicomedia and Theognis of Nicaea. Constantine was furious and exiled both bishops, ordering the people of Nicomedia and Nicaea to elect new bishops. He was especially incensed at Eusebius, whom he accused of having been a partisan of Licinius.

Constantine was particularly anxious that the deposed and exiled Arius should return to the flock and invited him to court. Arius came and submitted a statement of faith which avoided commitment to the *homoousion* but found acceptance with the emperor. Constantine asked Alexander to receive and reinstate Arius; Alexander refused. Constantine asked again, using stronger words; Alexander refused again. Constantine reconvened the Council of Nicaea (at Nicomedia, in 327), which readmitted Arius. At the same time Eusebius of Nicomedia and Theognis of Nicaea were recalled and restored to their sees. And Eusebius was soon to become Constantine's principal advisor on ecclesiastical matters, replacing Ossius, who had returned to his native Spain.

It was probably also in 327 that Eustathius, the bishop of Antioch and a staunch defender of orthodoxy, was driven from his see; his enemies had not shied from bringing all manner of false charges against him. Eusebius of Caesarea, who had played a central role in all of this, could have succeeded to the vacancy, but declined and earned Constantine's praise for so doing.

On 17 April 328 Bishop Alexander of Alexandria died and was succeeded, on 8 June, by his former deacon Athanasius, who refused even more adamantly to reinstate Arius, defying the emperor's orders. The new bishop of Alexandria was as strong-willed as the emperor. The dispute continued. In 332 Athanasius defended himself successfully before the emperor against various charges brought against him by his enemies. Another council met, on Constantine's orders, in 335 at Tyre under the supervision of an imperial commissioner, Flavius Dionysius. The council considered various charges brought against Athanasius,

50

found him guilty and deposed him from his bishopric. The council then interrupted its proceedings to attend, at Constantine's invitation, the dedication of the Church of the Holy Sepulchre at Jerusalem. Resuming their deliberations, the assembled bishops next considered the emperor's request to receive Arius and his followers into communion and acceded to the request. Arius, however, died in 336 in Constantinople, under less than dignified circumstances: while visiting a public latrine.

As for Athanasius, he decided to present his case in person to the emperor. Having left Tyre secretly and under cover of night, he arrived in Constantinople on 30 October 335. He accosted Constantine on one of the roads leading into the city. The emperor was taken by surprise but, impressed by Athanasius' personality, changed his mind, and severely rebuked the bishops who had passed judgment on Athanasius. Then several bishops, including both Eusebii, hurriedly came to Constantinople and managed to change the emperor's mind again; they said, among other things, that Athanasius had threatened to interfere with the shipment of grain from Alexandria to Constantinople. Athanasius was sent into exile at far-away Trier; he left Constantinople on 7 November, one week after he had arrived. He was released only after Constantine's death in 337, by an action of Constantine II. He left Trier equipped with a letter of introduction addressed by the new emperor (still Caesar, not yet Augustus) to the Church at Alexandria and dated 17 June. On his way back to Egypt he stopped at Viminacium in Moesia Superior for an audience with Constantius II. He also stopped in Constantinople and helped to consecrate a new bishop, Paul, who was to become a center of much controversy. Only on 23 November did Athanasius arrive in Alexandria.

Athanasius visited Trier again in 343 and perhaps also in 346. In any event his stays in the city left their mark on the church there: Bishops Maximinus and Paulinus of Trier were courageous defenders of the person and doctrine of Athanasius in the 340s and 350s.

51

8

The crisis in the imperial family

The years 324 and 325 saw Constantine at the pinnacle of success. His victories over Licinius had made him the sole ruler of the Roman world. He had brought unity to the Christian church, or so he thought, by convening the Council of Nicaea and taking an active part in it. He was observing his twentieth year in power. But tragedy struck in 326.

At some time in that year – attempts to arrive at a more specific date have not led to any conclusive result – Constantine ordered the execution of his son Crispus. The order was carried out at Pola in Istria. Was Crispus travelling by himself or was he accompanying his father? Was he on his way to Rome, perhaps, there to celebrate his father's *vicennalia*, or was he on the return journey? We do not know. More importantly, what had he done to deserve a sentence of death from his own father? There are no hints of a gradual estrangement between father and son. Crispus had held the consulship in 318, 321 and 324; he had performed well in the recent war against Licinius. Constantine had celebrated the birth of Crispus' child, Constantine's first grandchild, in 322. Nothing lends support to the suggestion that Crispus was plotting against his father. Nor is it likely that Constantine, the product of a casual union himself, acted out of concern for the rights of his younger, legitimate sons and for the principle of dynastic legitimacy. Crispus' illegitimate birth had not stood in the way of his advancement thus far; why should it now? And by removing him, while his other sons were all under ten years old, Constantine weakened rather than strengthened his own position.

In the same year, soon after the death of Crispus, Constantine also brought about the death of his wife Fausta, after a marriage of nineteen years. He did this, reportedly, by having her placed in an overheated bath, and some have imagined, without any good reason, that this transpired in the Imperial Baths of Trier. The two deaths follow so closely one upon the other that a connection between them appears most likely. And, indeed, the *Epitome* of Aurelius Victor (fourth century) does report such a connection:

> At the instigation of his wife Fausta, it is believed, Constantine ordered that his son Crispus be killed. Then, when his mother Helena, deeply grieving for her grandson, rebuked him, he killed his wife Fausta by placing her in a hot bath.

Our confidence in the veracity of this report is weakened by the phrase "it is believed." The means of Fausta's execution give us cause to wonder as well, especially since John Chrysostom (*c.* 347–407), the embattled patriarch of Constantinople, claims to know that Fausta was killed by exposure in the mountains. Zosimus, in the sixth century, offers a much embellished version of the story first found in the *Epitome*:

> When all the power devolved on Constantine alone he no longer hid the evil nature that was within him, but allowed himself to do all things as he pleased. . . . His son Crispus, who had been honoured with the rank of Caesar, as previously mentioned, came under suspicion of being involved with his stepmother Fausta; Constantine destroyed him without any regard to the laws of nature. When Constantine's mother Helena was disturbed by these events and was taking the loss of the young man very hard, Constantine, as if to console her, corrected one evil by an even greater evil; he ordered an unbearably hot bath to be prepared, had Fausta placed in it, and had her taken out only when she was dead.

It is impossible now to separate fact from gossip and to know with certainty what offences Crispus and Fausta had committed. Several laws issued by Constantine in 326 indicate that he was greatly concerned at that time with protecting the sanctity of marriage. Does that allow us to conclude that the offenses committed by Crispus and Fausta were of a sexual nature?

Both Crispus and Fausta suffered *damnatio memoriae*, that is, their names were removed from public records and inscriptions. Nor did Constantine's surviving sons ever rehabilitate the memory of their mother. In his *History of the Church*, in one of the editions produced before 326, Eusebius had spoken of Crispus with repulsive flattery, even implying that the relationship between Constantine and Crispus was comparable to the relationship between the Father and the Son in the Trinity. But in his *Life of Constantine* Crispus and Fausta are both passed over in silence.

And what of Helena's role in this family tragedy? It is to be noted that Helena is not accused by either report of having intrigued against Fausta; we are told only that she was disturbed by the death of Crispus and reproached Constantine on account of it. Various other sources do not mention her at all in this context. Thus we have no reason to shift the responsibility for Fausta's death from Constantine to Helena. But she may have felt that Constantine condemned Crispus rashly, and, indeed, there seem to have been no judicial procedures at all. It is also conceivable that there were jealousies and rivalries between mother-in-law and daughter-in-law. None of this would convict Helena in a court of law, but the suspicions remain.

That Constantine became a Christian because Christianity offered him forgiveness for the sins which he had committed against his own kin was asserted by Julian the Apostate and repeated, predictably, by Zosimus. But any such notion is to be resolutely rejected. The crucial year for Constantine's conversion was 312, not 326. And, if he had really been troubled by the burden of sins which he was carrying, would he not have sought the forgiving sacrament of baptism much sooner than he did?

When she was already "advanced in years" Helena undertook a pilgrimage to the Holy Land, traveling not as a private person but as the representative of her son and as Augusta. We do not know when she set out on her pilgrimage, how many months or years she spent traveling, or when she returned. We do know that her visit took place after the Council of Nicaea and that she died shortly after her return, and numismatic evidence points to her death in 329. A journey of two years' duration, 326–8,

54

would thus seem reasonable. We do not know her precise itinerary either, but it is likely that she passed through Syrian Antioch; her presence in the area seems to have contributed to the downfall of Bishop Eustathius (see Chapter 7), who unwisely made an uncomplimentary remark about her. Numerous scholars have linked this pilgrimage to the tragedy which occurred in the imperial family in 326, and rightly so; it has even been suggested that it was an act of expiation, either for sins of her own or for sins of her son. Certainly it was an act of personal piety as well as a state visit. Eusebius praises her piety, humility and charity, and there is no reason to doubt him on this point. Constantine and Helena were united in a common pious purpose: the construction and embellishment of churches at Christendom's most holy places. These we must consider next.

When the emperor Hadrian after the Second Jewish War converted the Jewish city of Jerusalem into the Roman city of Aelia Capitolina, a temple of Venus was built on a platform which was piled up on the very spot which Christians believed to be the site of the Resurrection. After the Council of Nicaea, Constantine ordered the removal of the offending pagan temple and even of the platform on which it stood. In the course of the work the tomb which was believed to be – and in all likelihood is – the tomb of Jesus was discovered. Constantine then ordered the construction of a splendid church, the Church of the Holy Sepulchre, sometimes also called the "New Jerusalem." The church was dedicated on 13 September 335 by an assembly of bishops who were then attending the Council of Tyre (see Chapter 7). The dedication was part of Constantine's *tricennalia* and thereafter commemorated annually by the feast of the Encaenia. The church took the form of a five-aisled basilica, also known as the Martyrium. In a court adjoining the basilica at its west end there were the rock of Calvary or Golgotha and the sacred tomb, which was some years later enclosed by the famous rotunda known as the Anastasis ("Resurrection"), measuring one hundred feet in diameter. Our principal source of information is Eusebius, who credits Constantine only, without mentioning Helena. Constantine's building suffered many vicissitudes through the centuries, and the present structure, a Crusader church of the twelfth century, bears little resemblance to it.

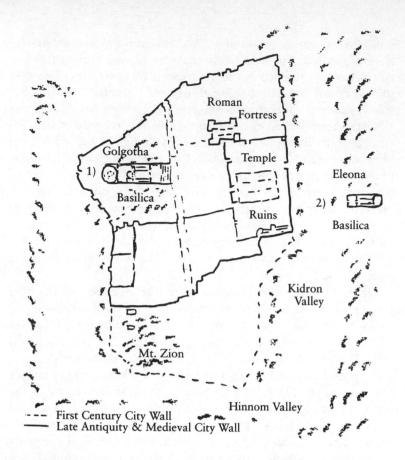

Roman Fortress

Golgotha

1)

Temple

Eleona

Basilica

2) ☐▭

Ruins

Basilica

Kidron Valley

Mt. Zion

Hinnom Valley

--- First Century City Wall
——— Late Antiquity & Medieval City Wall

1) Golgotha Basilica: the 130 meter Holy Sepulchre complex rose westward from the north-south cardo of Jerusalem with a marble staircase, an open courtyard, the basilica proper (Egeria's Sancta Ecclesia Martyrium), a second courtyard with a portion of Golgotha rock, and the shrine around Christ's tomb (Egeria's Sancta Ecclesia Anastasis).
2) Eleona Basilica: this complex of *c.* 70 meters rose eastward up the Mt. of Olives with a courtyard, a basilican church, and a shrine over a cave where Jesus had taught.

Figure 6 Map of the Christian basilicas of Constantinian Jerusalem. From Charles M. Odahl, *Early Christian Latin Literature* (Chicago 1993), p. 157.

Courtesy of Charles M. Odahl and Ares Publishers, Inc.

At Bethlehem the Church of the Nativity was built over and around the cave or grotto which was believed to be the place where Jesus was born. Again Eusebius is our principal source of information, but his imprecise language does not make clear what specific roles Constantine and Helena played in the construction of this church; a joint endeavor of the emperor and his mother seems reasonable. This church was similar in plan to the Church of the Holy Sepulchre, but of somewhat smaller proportions. In the sixth century the emperor Justinian replaced the original structure with a slightly larger one, and it is Justinian's church which stands to this day; the floor mosaics of the original fourth-century church have been found in archaeological explorations a couple of feet below the later floor.

A third church to interest us in the present context is the Church of the Eleona, or Church on the Mount of Olives. Again Eusebius fails to make clear specifically how Constantine and Helena respectively were involved in the establishment of this church, and again we may suppose a joint endeavor. This church was more modest in size and a three-aisled basilica in shape. At the eastern end of the basilica there was an apse, and below this apse was a crypt or cave, in which Christ was believed to have taught his disciples. Only the foundation trenches, small segments of the foundation walls, and parts of the floor mosaics of the nave remain today.

Later accounts, not very credible, credit Helena with having founded twenty-eight, thirty or forty-eight churches in the course of her pilgrimage. She is supposed to have visited Mount Sinai, Aleppo and Cyprus. But the most glorious achievement of Helena's pilgrimage was the supposed *inventio* (discovery) of the True Cross of Christ's Passion. Most scholars assign this achievement, for good reasons, to the realm of legend rather than of historical fact. It accounts, nevertheless, for one of Christendom's most cherished traditions, one which has found rich expression in hagiography and in hundreds of works of sacred art. It also earned Helena the rank of saint. (Her feast day in the West is 18 August, while in the East she shares a feast day, 21 May, with her son. The Exaltation of the Cross is observed in both the West and the East on 14 September.) Quite apart from the claims of legend, her pilgrimage was clearly a memorable event. But Helena was not the first pilgrim to the Holy Land, although this has at times been asserted.

The fourth Constantinian church in the Holy Land is associated not with Helena but with Eutropia, the emperor's mother-in-law. This lady, whom Constantine seems to have held in high regard, was traveling in the Holy Land at about the same time as Helena, but nothing suggests that they traveled together. Among the places visited by Eutropia was Mamre (near Hebron), a holy place to Jews and Christians because here, according to Genesis 18, Abraham had hospitably entertained three divine messengers in the shade of an oak tree or terebinth. The lady found that this holy place was defiled by pagan rites as well as by secular activities. She reported to Constantine, who at once ordered the place to be purified and a church to be built there. The church was completed by 333. Remnants of the outer walls exist to this day, and we know that the atrium contained Abraham's altar, the well and the tree. Remnants of that tree were still seen by St Jerome.

Constantine's building activities in Palestine gave to that country a central place in Christian sentiments which it had never had before; they in fact made it the "Holy Land." We shall examine an even more ambitious project of the emperor in the next chapter.

9

The new Rome

In the second half of the seventh century BC Greek colonists from
Megara founded the city of Byzantium on the European shore
of the Bosporus. In about 150 BC, even before Macedonia
became a Roman province (148 BC), this city became a de-
pendency of the Romans. It soon became the eastern terminus
of the Egnatian Way, which led to Thessalonike in Macedonia
and to Dyrrhachium on the Adriatic. The emperor Septimius
Severus laid siege to the city, which had sided with his rival
Pescennius Niger, in AD 193, took it after almost three years,
destroyed its walls, and let his soldiers loot it. He soon rebuilt
it on a larger scale. Constantine, too, as mentioned in Chapter
6, had to besiege the city in the course of his second war against
Licinius. The advantages of its location will not have been lost
to Constantine's trained eye, just as they had not been lost to
the Athenians or to the Spartans in the Peloponnesian War, or
to Philip II of Macedon in his quest for power.

Byzantium was indeed favorably located. It occupied a key
position at the crossroads between Asia and Europe, linking east
and west and north and south. This location afforded easy
access to the Balkan provinces, which played such an important
role in the third and fourth centuries and of which Constantine
was fond. From here, too, the eastern frontier, which so often
claimed the emperors' attention, could more readily be reached
than it could from any of the western residences. Anyone who

59

was master of the city was also in control of the traffic of ships through the Bosporus to and from the Euxine (the Black Sea). It is easy to see why the location has been of strategic significance through the centuries.

Byzantium was situated on a triangular piece of territory which was bordered on the west by Thrace, on the south by the Propontis (the Sea of Marmara) and on the north by an inlet, later known as the Golden Horn, which provided a natural deep-water port. The land of Thrace was fertile, the climate pleasant, and the waters of the Propontis rich in fish. All the conditions for a favorable economic development were met. Edward Gibbon thought that Constantinople "appears to have been formed by Nature for the centre and capital of a great monarchy." Across the Bosporus, on the Asiatic side, there were the cities of Chrysopolis and Chalcedon, and c. fifty miles further east lay Nicomedia, which had served Diocletian as his residence.

Long before Constantine various Greek kings and Roman emperors had founded or refounded cities and then named them after themselves; Philippopolis and Adrianople may serve as examples. The tetrarchs had not resided in Rome; neither had Constantine chosen to do so after his victory over Maxentius. But Constantine went further than that. His new city would challenge the primacy of Rome as none of the imperial residences had hitherto done. His decision to refound Byzantium as Constantinople and to make it the capital of his empire was a momentous one; it would affect the course of world history for centuries to come; it has rightly been compared with the founding of Alexandria centuries earlier or with that of St Petersburg centuries later. This decision ranks in importance and profundity with his decision to adopt Christianity. The new city became a center of Christianity, the see of a patriarch, comparable in stature to Rome, Alexandria, Antioch, or Jerusalem. As the "New Rome" the city inherited the political institutions of the old Rome, but it also inherited the cultural traditions of the Greek East.

This momentous decision must have slowly ripened in Constantine's mind. The building of the family mausoleum on the Via Labicana just outside Rome clearly indicates that at one time he must have meant for himself to be buried there. One can imagine that there was a gradually increasing estrangement between the people of Rome and the emperor. The city, filled with pagan buildings and institutions, must have made Constantine uncomfortable. He must have observed with chagrin that many members of the senatorial aristocracy clung to their pagan ways; the altar and statue of Victory still stood in the senate house. The people of Rome, in turn, took offence at an emperor who pointedly omitted the customary sacrifices when he entered the city as a victor in 312, when he observed his *decennalia* in 315 and when he observed his *vicennalia* in 326. Rome would not become the Christian capital of Constantine's empire. Political and strategic considerations dictated that the new, Christian capital should be in the East, whither the empire's center of gravity had shifted.

On 8 November 324 Constantine appointed his son Constantius to the rank of Caesar. It has been suggested, but not established, that on this day also he bestowed the rank of Augusta on his wife Fausta and his mother Helena. More important in the present context is the fact that on this day – a Sunday – he formally laid out the boundaries of his new city, moving them *c.* four kilometers further out, and roughly quadrupling its territory. If we may believe the report of the fifth-century historian Philostorgius, Constantine traced the line of the future walls on the ground with a spear, in the manner of a Greek *ktistes* (founder). The same historian reports that Constantine's companions were amazed at the vast circumference of the new walls and Constantine responded, "I shall keep on until he who walks ahead of me will stop." Was he experiencing another vision? In his own mind, was his action on that day as significant as his victory at the Milvian Bridge? Surely Constantine would have been disappointed if he had known that less than a century later the walls of Theodosius II doubled the territory of the city once again and that today it is the walls of Theodosius which stand, not his own.

The building and settlement of the new city were carried forward with great speed. The new walls were completed by 328. The emperor offered various incentives to people to settle in his new city, especially if they were skilled in the building trades. The technical and logistical challenges must have been extraordinary, and the achievement was indeed remarkable. On 11 May 330 the new city was formally dedicated with rites taking place in the hippodrome. The pagan augur Sopater (later put to death by Constantine) had chosen the day as propitious. Coins minted that year announced the event to the world; on the obverse of these coins the figure of Constantinopolis carried a cross scepter over her shoulder, thus emphasizing the Christian character of the city. The festivities lasted forty days. On the last day of the festivities a golden statue of the Tyche (Fortune) of Constantinople was paraded through the streets. This was hardly a Christian practice, but it should not be seen as a return to paganism on Constantine's part: he, and most of the people, probably thought of Tyche not as a pagan goddess but as an abstraction or a personification of the city. A home for the Tyche was found in a former shrine of Cybele. And 11 May was henceforth observed as a holiday, just as Rome observed 21 April as its birthday.

The new city was modeled on the city of Rome in more than one way (Figure 7). Like the old capital, it was was built on seven hills and divided into fourteen administrative districts (although two of these were ouside the city walls). There was a senate, just as in Rome; its members, however, ranked below the members of the senate in Rome, being styled *clari* (distinguished) rather than *clarissimi* (most distinguished), and there were two senate houses rather than one. The people of Constantinople received subsidized grain, just as the people of Rome did. We have seen in Chapter 7 above how readily Constantine was moved by the report that Athanasius had threatened to interfere with the shipment of grain from Egypt to Constantinople.

More than Trier, or Rome, or Jerusalem, Constantinople gave to Constantine an unparalleled opportunity for building and planning on a grand scale. Constantine did not, however, have to create a completely new plan. Rather, he retained certain features of the Severan city, such as the *Mese* (the principal avenue), the agora, which became the Augusteum (named after his mother, the Augusta), and the location of the hippodrome. He enlarged and embellished the existing Baths of Zeuxippos. Perhaps he found in place already the so-called Column of the Goths. This column possibly celebrates a victory won by Claudius Gothicus, Constantine's supposed ancestor, over the Goths in 269 at Naissus, Constantine's birthplace. The inscription on the base of the column reads, *Fortunae reduci ob devictos Gothos*, "to returning Fortune on account of the defeat

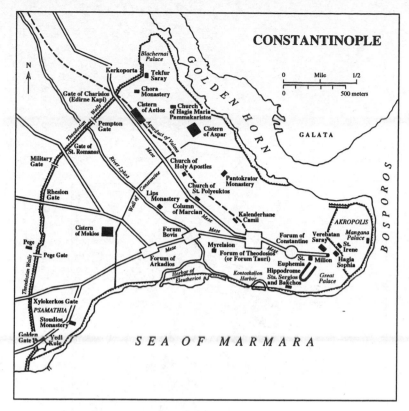

Figure 7 Map of Constantinople.
Courtesy of Dumbarton Oaks, Washington, DC.

of the Goths." Unfortunately it does not record when or by whom the column was erected. Thus it is possible also that the column was erected later by Constantine himself to celebrate a victory of his own over the Goths.

In the area now occupied by the Mosque of Sultan Ahmet (the Blue Mosque) Constantine built the imperial palace (Figure 8). The palace afforded direct access to the *kathisma*, the royal box overlooking the hippodrome, and probably also to the Church of Hagia Sophia. The elaborate entrance to the palace was known as the *Chalke* (Brazen) Gate. A painting over the portal, described for us by Eusebius, depicted a victorious Constantine

Figure 8 The Great Palace of Constantinople. From George M. A. Hanfmann, *From Croesus to Constantine* (Ann Arbor, 1975), Fig. 176.

Courtesy of The University of Michigan Press.

in the act of slaying a dragon; did the dragon represent Licinius, paganism or, more generally, all the forces of evil?

The hippodrome, which Septimius Severus had begun but which was still unfinished when Constantine became master of the city, was completed and enlarged to a capacity of *c.* 50,000. In the center of the *spina* (median) the emperor placed the Serpent Column from Delphi. This column had been erected to commemorate the victory which the Greeks had won in 479 BC at Plataea over the Persians. It was only one of many objects which Constantine appropriated for the embellishment of his city, causing St Jerome to remark that "nearly all cities were stripped bare." The column stands in Constantinople to this day, in an enclosure, its bottom considerably below the level of the modern street. The golden tripod which once crowned it had been lost already in the fourth century BC; the circular base of the column is still to be seen in Delphi today. At the south end of the *spina* Constantine erected a tall obelisk, which was restored by Constantine VII Porphyrogenitus in the tenth century. More attention is usually given to the great Egyptian obelisk which, after a delay of several decades, was finally put into place near the other end of the *spina* by Theodosius II; it had been one of a pair of obelisks erected by Thutmose III at Karnak. The hippodrome served not only for the entertainment of the people, but also for public acts of state.

In front of the Old Gate of the Severan walls Constantine built the forum which bore his name. It was circular or elliptical in shape, thus recalling the oval forum of Gerasa (Jerash) in Jordan. It was enclosed by a double tier of colonnades, which was interrupted by two arched passageways providing access from two major streets, the *Mese* and the *Regia*. In the center of the open space stood the Column of Constantine. Built of nine drums of porphyry, it was 25 m high, while its diameter measures 2.90 m. At one time it was heavily damaged by fire; what remains of it today is enclosed by iron bands and called the Burnt Column. The base of the column was enclosed in a tetrapylon, which was provided with an altar, so that Mass could be celebrated there. The column was crowned by a statue of Helios, its features suitably adapted so as to represent Constantine. The figure wore a crown of seven rays; in the left hand it held a spear, in the right hand a globe. The historian Socrates, in the fifth century, claims to know that Constantine

65

deposited in the statue a part of the True Cross, sent to him by his mother. Gregory of Tours, in the sixth century, has heard that one of four Holy Nails found by Helena was put in the head of the statue. It is easy to dismiss these reports as the legends which they are. It is more difficult to interpret the statue. Did Constantine think of himself as an epiphany of Helios, or did he want his subjects to think of him thus? And did he associate, in his mind, Helios with Christ? The mosaic of Christ-Helios in the mausoleum of the Julii, in the Vatican necropolis, deep under the floor of St Peter's Basilica, suggests that such association did indeed occur. The logical, but blasphemous, conclusion of such association then would be that Constantine thought of himself as an epiphany of Christ. The altar at the foot of the statue would lend support to such conclusion. Or had Constantine not thought this far? Did he "merely" think of himself as Christ's agent, ruling on earth as Christ rules in heaven? Perhaps he had read Romans 13:1, "Let every person be subject to the governing authorities. For there is no authority except from God, and those that exist have been instituted by God" (Revised Standard Version). Interestingly enough, Eusebius is silent about the column, the statue, and the altar.

A less controversial monument, part of which survives, was the Milion, a small structure, probably a tetrapylon, from which all distances were measured. It was comparable to the Miliarium Aureum in the Forum of Rome. It gave visible expression to the city's actual and claimed centrality. Two Byzantine texts, of the eighth and tenth centuries respectively, report that the Milion was topped by statues of Constantine and Helena, with a cross between them. If these reports are true, it must nevertheless be said that these statues cannot have been of Constantinian date, since they are evidently inspired by the post-Constantinian legend of the True Cross.

Constantine began the construction of two major churches in Constantinople, the Church of Hagia Sophia (Holy Wisdom) and the Church of Hagia Eirene (Holy Peace). These two designations would not be offensive to pagans. Was Constantine, even now, and even in his Christian capital, practising a kind of neutralism? He did not see either church completed in his lifetime. Work on Hagia Sophia, also known as "the Great Church," was begun in 326, according to one of our sources, but the dedication did not take place until 15 February 360, under Constantius II. Thirty-four years of construction seems inordinately long, considering that only ten years were required for the completion of the Church of the Holy Sepulchre in Jerusalem; an element of uncertainty is thus introduced. The church was severely damaged by fire in 404 and re-consecrated by Theodosius II in 415. It was destroyed in the Nika riots of 532 and replaced by Justinian with the church which stands today as the foremost monument of Byzantine architecture. Of the original Constantinian church we know that it was a five-aisled basilica with galleries, an atrium in the west, and an apse in the east. The original Hagia Eirene, in close proximity to Hagia Sophia, also fell victim to the Nika riots of 532. The replacement church, erected by Justinian and remodeled in the eighth century after an earthquake, stands to this day, but bears little resemblance to the original.

Concerning the foundation of the Church of the Holy Apostles there are two traditions in our sources and in modern scholarship. One of these holds that the church was built by Constantine, or at least begun by him; the other holds that it was built by Constantius II. This writer has been persuaded by the report of Eusebius and by the expert opinions of Professors Richard Krautheimer and Gilbert Dagron that the former tradition is the correct one. We shall consider this Constantinian foundation further in the next chapter, in the context of Constantine's burial.

The three churches here mentioned, Hagia Sophia, Hagia Eirene and Holy Apostles, were not only the largest in Constantinople, but also the most important, both politically and ecclesiastically. The Church of the Holy Apostles occupied the highest hill in the city (the fourth) and thus was visible from the Bosporus, while the other two were in immediate proximity to the palace.

Constantine's architects, in both the West and the East, created the first monumental Christian buildings, as distinguished from the modest *domus ecclesiae* which had served as Christian meeting places before Constantinian times. They employed both longitudinal and central designs, the former to meet liturgical needs, the latter to serve memorial purposes. They wrote the first chapter in the history of ecclesiastical architecture.

10

Constantine's government

We are not as well informed on the secular aspects of Constantine's reign as we would like to be; our sources and many historians after them seem to have been more keenly interested in the religious aspects. But some observations may be made.

In general Constantine refrained from sweeping innovations, being content, for the most part, with completing or continuing the arrangements made by Diocletian. One notable change pertained to the praetorian prefects: these now became civilian ministers. The Augustus and each of the Caesars were assisted by a praetorian prefect; so, for instance, Crispus, when at a very young age he was put nominally in charge of the western provinces. In a rather unusual arrangement a praetorian prefect administered the province of Africa. At the next lower level each diocese, normally comprising three provinces, was headed by a *vicarius*. The provincial governors held one of two ranks; either the higher rank of *consularis* or the lower rank of *praeses*. The city of Rome remained apart from the provincial organization and was administered by the urban prefect.

In the civil administration of the empire there seems to have been much corruption, and Constantine thundered against it: he threatened to "cut off the rapacious hands of (corrupt) officials" or to "sever the heads and necks of the villains." The many rescripts (legal decisions) issued by him or his staff reveal a concern with the minutiae of administration.

Some of Constantine's measures show a genuine concern for the welfare of his subjects. On one occasion he ordered the distribution of money, food and clothing to poor parents in Italy and Africa. When a famine struck Syria in 334 he had food supplies distributed through the churches.

Constantine, like a Roman censor of Republican times, was anxious to protect the moral fibre of Roman society, especially in matters of sexual conduct. He even ordered that parents who had been accessory to the seduction of a daughter should be punished by having molten lead poured down their throats! If a free woman and a male slave were found to have had sexual relations they were both to be put to death, the slave by being burned alive. A number of rescripts meant to protect the sanctity of marriage were issued in 326; unfortunately we do not know how these relate to the crisis which occurred in the imperial family in that year (see Chapter 8).

Gladiatorial shows were outlawed in 325, but continued in the West until the beginning of the next century. Crucifixion as a means of execution was also outlawed. And criminals were not to be branded in the face, not because Constantine objected to the cruelty, but because the face is "shaped in the likeness of heavenly beauty." Those confined to prisons were not to be deprived of daylight all together. The exposure of infants was prohibited.

Slaves benefited from Constantine's legislation in at least two ways. He provided that slave families were not to be separated when an estate was broken up; and he allowed that manumission of slaves could take place in church. On the other hand, if a slave had been been flogged by his master so severely that he subsequently died no charges were to be brought against the master. And slavery as an institution was never questioned.

Constantine spent money lavishly on benefits to the church, on building projects and on generous gifts to individuals or groups. He generated additional revenues by instituting two new taxes: one on the landed property of senators and the other on the tradesmen in the cities; the latter, oddly enough, was assessed every fifth year. The confiscation of pagan temple treasuries probably brought in less money than the emperor's critics asserted.

In the area of currency reform Constantine enjoyed more success than Diocletian had (see Chapter 2). A new type of coin, the gold *solidus*, was particularly successful. It won acceptance even beyond the borders of the empire and remained undebased until the eleventh century. A sixth-century writer admiringly remarks that "all nations trade in their (the Romans') currency and in every place from one end of the world to the other it is acceptable and envied by every man and every kingdom." Constantine's bronze coinage was less successful.

And then there was the question of succession (Figure 9). Constantius I and Theodora had had three sons and three daughters: Flavius Dalmatius (appointed "censor" in 333), Julius Constantius (appointed consul in 335; father of Julian the Apostate), and Hannibalianus; Constantia (see Chapters 4 and 6), Anastasia (the Christian connotations of her name raise interesting questions), and Eutropia (not to be confused with Constntine's like-named mother-in-law). Flavius Dalmatius had a like-named son, whom Constantine appointed Caesar on 18 September 335. Constantine's sons had been appointed to that rank previously: Constantine II on 1 March 317 (see Chapter 6), Constantius II on 8 November 324 (see Chapter 9) and last Constans on 25 December 333. Thus, during the last two years of Constantine's reign, there were, once more, four Caesars. Constantine II resided at Trier, Constans probably at Milan, Flavius Dalmatius probably at Naissus, and Constantius II at Antioch; each Caesar had a praetorian prefect by his side. The Augustus maintained his principal residence at Constantinople. Additionally, the younger Hannibalianus, another son of the elder Flavius Dalmatius, was appointed to the curious post of "king of kings and of the Pontic nations," evidently after a Persian attack upon Armenia; he also was given Constantina, Constantine's daughter, in marriage. It is not clear which of the Caesars Constantine intended to take precedence upon his own death.

Constantine continued and further developed an organization of the army which had first been instituted by Diocletian: the army was divided into two branches, the border troops, the *limitanei* and *ripenses*, and the mobile field army, called the *comitatenses*. In each area of defence the border troops were commanded by a *dux*, not by the provincial governor. The field army, which was comprised of elite units and enjoyed superior privileges, was under the command of the emperor himself or one of the Caesars, assisted by a Master of the Infantry and a

A. Constantius I and Helena

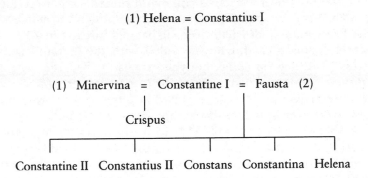

B. Constantius I and Theodora

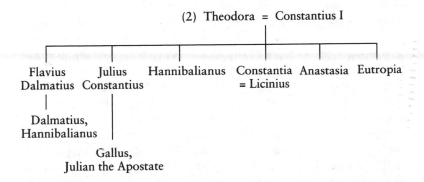

Notes:
1. Fausta was the daughter of Maximian and the sister of Maxentius.
2. Helena, daughter of Constantine, married Julian the Apostate.
3. Theodora was the daughter (or stepdaughter) of Maximian.

Figure 9 Genealogical tables.

71

Master of the Cavalry. The provincial governors and the praetorian prefects were relieved of most of their military functions; the latter no longer exercised command functions, but retained responsibility for providing recruits, rations and armaments. The Praetorian Guard had been disbanded already in 312 – a suitable punishment for having sided with the "tyrant" Maxentius. The new imperial bodyguard was called the *scholae palatinae* and was commanded by the *magister officiorum*. Increasingly Germans were recruited into the Roman army, and some were even appointed to high military command.

In the years 306–8 and 314–15 Constantine campaigned successfully on the German frontier, as we have seen in Chapters 3 and 5. The security of the empire claimed his attention again in his later years. He campaigned with some success on the Danube frontier against the Goths in 332 and against the Sarmatians in 334. His campaigns resulted in the partial and temporary recovery of Dacia, which the emperor Aurelian (270–5) had abandoned; Constantine now added *Dacicus Maximus* to his titles. On Cyprus a certain Calocaerus led an uprising which is best dated to 333–4. This uprising was suppressed by Flavius Dalmatius, Constantine's half-brother (see above), and the would-be usurper (a camel-driver by background!) captured and cruelly put to death at Tarsus. In 336 the emperor was preparing for a campaign against the Persians; a Persian embassy in the winter of 336–7 was rebuffed.

Constantine, like Diocletian before him, surrounded himself with an elaborate court ceremonial. He and everything associated with him became sacred. In his presence a respectful silence was expected; hence the ushers were called *silentiarii*. His advisors stood when meeting with him in council; hence they became known as the *consistorium*. The elaborate remoteness of the emperor not only protected him against plots; it also isolated him from his subjects. Nevertheless Constantine encouraged them to address their petitions to him.

The prevailing spirit of Constantine's government was one of conservatism. His conversion to and support of Christianity produced fewer innovations than one might have expected; indeed they served an entirely conservative end, the preservation and continuation of the empire. There is an element of continuity also in the founding of the New Rome, which in more than one way was modeled upon the old Rome.

11

Constantine's final years, death and burial

In the years between 325 and 337 Constantine continued his support of the church. In a letter to Eusebius of Caesarea he noted with pleasure that the number of Christian converts in Constantinople had grown greatly and that there was a need to build additional churches. He then requested Eusebius to oversee the production of fifty exemplars of the Holy Scriptures and their conveyance by public transport to Constantinople. Eusebius undertook the work at once, seeing to it that these Scriptures were gathered in magnificent and elaborately bound volumes. Constantine acknowledged the receipt of these volumes and at the same time expressed his joy at the growth of the church in Constantia in Palestine (the former Maiuma, the port of Gaza, renamed in honor of his deceased sister Constantia). Constantine also supported Christian charities by granting allocations of grain to churches for distribution to the poor. When the city of Antioch was struck by famine in 334 grain was distributed to the people through the churches.

And Constantine continued to apply the resources of the state to the building of churches. At Aquileia a double basilica was built in Constantinian times; only the south church still stands. In two of the ten panels of a mosaic floor which has been dated to 325 we find a series of portraits which possibly depict Constantine and members of the imperial family. At Nicomedia the emperor ordered the building of a "stately and magnificent

73

church." At Antioch the great Octagon, or "Golden Church," "a church of unparalleled size and beauty," was begun in 327 and dedicated in 341. At Cirta, the capital of Numidia, renamed Constantina in the emperor's honor, Constantine paid for the construction of a basilica, which was completed in 329. When it was seized by the Donatists, another one for the Catholics was promptly begun, again at the emperor's expense (330).

The emperor's championship of the Christian cause also became a factor in foreign relations. The small kingdom of Iberia (today Georgia) in the Caucasus adopted Christianity at some time during Constantine's reign, but it cannot be shown that this was due to any action on Constantine's part. We are better informed about the situation of Christians in Persia. There, after a period of persecution in the late third century, the Christians were tolerated until they came under suspicion, not unfounded, of harbouring pro-Roman sentiments. Among the bishops assembled at Nicaea in 325 there was one from Persia. In a letter to the Persian king Shapur (or Sapor) II in 325 or 326, Constantine represented himself as the protector of Christians in Persia. In Armenia, neighbouring both the Roman empire and the Persian kingdom, King Tiridates (Trdat) III (287–330) had been converted to Christianity by Saint Gregory the Illuminator, and his kingdom had officially become Christian early in the fourth century (there is no consensus on the specific date). According to an Armenian tradition the king and the saint, now a bishop, traveled, some years later, to Constantine's court and were received with the highest honors.

But Constantine went beyond benevolent sponsorship of the Christian cause; he actively suppressed paganism, at least in some specific instances. The construction of the Church of the Holy Sepulchre at Jerusalem and of the basilica at Mamre involved the destruction of pre-existing pagan shrines at these sites, as we have seen. At Aphaca (Afqua) on Mount Lebanon, at the headwaters of the Adonis River (today's Nahr Ibrahim), traditionally the site of the myth of Aphrodite and Adonis, the cult of Aphrodite reportedly involved much licentiousness; Constantine had the temple torn down. At Heliopolis (Baalbek) in Phoenicia he forbade the cult of Venus Heliopolitana, which included ritual prostitution, and ordered the construction of a church. At Aegeae in Cilicia the emperor's soldiers tore down the temple of Asclepius, the center of a popular healing cult. At

Antioch the Temple of the Muses was diverted to secular purposes. Elsewhere temple treasures were confiscated and the proceeds fed into the imperial treasury. Far from censuring such forceful means, Eusebius, in his *Panegyric to Constantine*, on 25 July 336, commended the emperor for having "cleansed all the filth of godless error from his kingdom on earth." But, in spite of what Eusebius reports, there is no evidence of a general cessation of pagan rites throughout the empire.

We have seen already that Constantine harbored strong anti-Jewish sentiments. These sentiments were translated into anti-Jewish legislation which is best assigned to Constantine's later years, although the specific dating is problematic. Constantine allowed the Jews access to the city of Jerusalem once a year – to bewail their fate; the city had been off limits to Jews since Hadrian's founding of Aelia Capitolina. Constantine provided that Jews who attempted by force to prevent the conversion of a co-religionist to Christianity were to be burned alive! Jews were not allowed to own Christian slaves, to circumcise any male slave in their possession or to make converts to their faith. On the other hand a certain Josephus of Tiberias, a convert from Judaism to Christianity, was elevated to the rank of *comes* (count) and given permission and money to build churches in Galilee.

We now return to the subject of Constantine's baptism. We have already rejected in Chapter 4 the notion that he was baptized in 312 by Pope Sylvester in Rome. Let us now examine the circumstances which really surrounded his baptism and his death.

Shortly after Easter (3 April) of 337 Constantine felt the onset of illness. He repaired to Drepanum = Helenopolis and there, in a final expression of filial piety, prayed at the tomb of the martyr Lucian, his mother's favorite saint. He then proceeded to the suburbs of Nicomedia and there summoned "the bishops", to use Eusebius' curious phrase. Addressing them, he explained that it had been his fervent hope to be baptized in the Jordan River; but now it was his desire to receive the saving sacrament right then and there. To use Eusebius' language again, "The prelates performed the sacred ceremonies in the usual manner and . . . made him a partaker of the mystic ordinance." Eusebius was a master at evasion; he does not identify the bishop who performed the baptism. The *Chronicle* of St Jerome, more than

75

forty years later, provides the missing information: "At the end of his life Constantine was baptized by Eusebius, the bishop of Nicomedia, and fell into the Arian doctrine." Ruefully, St Jerome adds "Hence there has followed to this day the decline of the churches and discord over the whole earth." The *Chronica* of Bishop Isidore of Seville, in the seventh century, also expresses regret: "What sorrow! He made a good start but a bad finish." The embarrassment felt by Jerome and Isidore is understandable: how much more appropriate it would have been for the ruler of the empire to receive baptism in Rome at the hands of the Pope! For the remaining days of his life Constantine did not dress in the imperial robes but in the white robes of a Christian neophyte.

On the day of Pentecost, 22 May, in 337, Constantine died at Nicomedia; his body was escorted to Constantinople and lay in state in the imperial palace. He had provided with forethought for his own burial: his sarcophagus was placed under the central dome of the Church of the Holy Apostles, surrounded by the cenotaphs or memorial steles of the Twelve Apostles, making him symbolically the thirteenth Apostle. No one raised an objection at the time to this extraordinary arrangement, but in 359 Constantius II had the sarcophagus removed to the Church of St Akakios. In 370 it was returned to the Church of the Holy Apostles, not, however, to its original location, but rather to a recently constructed circular mausoleum which was attached to, but separate from, the church – a far more acceptable arrangement – and which was to provide space for the burial of future emperors as well. Of interest also is that in 356 and 357 the church received relics of Timothy, the companion of Paul, of the Apostle Andrew and of the Evangelist Luke. The deposition of these relics changed the nature of the church and may have prompted the removal of Constantine's sarcophagus. Later Byzantine authors will have us believe that the first member of the imperial family entombed in the Church of the Holy Apostles was the emperor's mother Helena. Some of them even claim that the emperor and his mother shared a single sarcophagus. These reports are without any basis in fact. The Constantinian foundation was replaced in 550 by Justinian with a church of his own. Today the Fatih Mosque, or Mosque of Mehmet the Conqueror, built in 1461–73, stands on the site once occupied by the Church of the Holy Apostles; it incorporates in its fabric some beautiful columns taken from the earlier Christian structure.

After Constantine's death and funeral it remained unclear for some time who would hold supreme power. Constantius had hastened to Constantinople from Antioch and, in the absence of his brothers, had supervised the funeral, but he did not proclaim himself Augustus. It was apparently the army which decided that the succession should be limited to Constantine's surviving three sons. In a bloody coup possible rivals were eliminated: the elder Dalmatius and Julius Constantius, both the late emperor's half-brothers; also his nephews, the Caesar Dalmatius and the younger Hannibalianus, the "king of kings." The elder Hannibalianus, Constantine's third half-brother, had apparently died at an earlier time. Two young sons of Julius Constantius, Gallus and Julian, were permitted to live. (The former was appointed Caesar in 351 but executed in 354 under suspicion of treason. The latter assumed his place in history as Julian the Apostate.) Ablabius, Constantine's right-hand man and praetorian prefect to Constantius II, was another victim of the turbulence of this year. Laws continued to be issued in the name of the deceased emperor. Only on 9 September 337 did Constantine II, Constantius II and Constans each assume the title of Augustus.

12

Constantine's image in Roman art

The portraits of the Roman emperors must be studied not only from an aesthetic perspective; rather they must be examined also for the ideological and political message which they are meant to convey. Thus the members of the First Tetrarchy are regularly portrayed with a stern, even frightening countenance. But even Constantine's earliest portraits, on coins minted at Trier in 306, show a departure from that model by a softer, more natural modeling and by a classicizing tendency.

This development manifests itself more clearly in the medallions of Constantine on his arch in Rome. Although these medallions are recut spoils from monuments of Trajan and Hadrian they have a rather good claim on authenticity. They show a handsome and youthful emperor, with longer hair, and clean-shaven, in a break with previous tradition. A narrow band of relief sculptures, in six panels, beginning on the narrow west face of the arch, celebrates the events of the year 312: the departure of Constantine's army from Milan, the siege of Verona, the victory at the Milvian Bridge, the entry into Rome, the address to the people, and the distribution of donatives. Of interest also is that a place is found on the arch for a portrait of Claudius Gothicus, Constantine's supposed ancestor. It is true that the arch was erected not *by* Constantine but *for* Constantine by the senate; nevertheless it appears that it served

78

his "public relations programme" and that his desires were taken into account.

The silver medallion, minted at Ticinum in 315, showing Constantine with the Chi-Rho monogram on his helmet, has already been mentioned in Chapter 6.

A new look is to be noted on Constantine's coins from 324 on. His gaze is now turned upward and into the distance; this reflects his claim not on divinity but on divine mission and inspiration. From this time on, too, he stops wearing a laurel wreath and adopts the diadem. The precedent for this is to be sought not in a Christian but in a Hellenistic context, in Alexander the Great. It is not by coincidence that Eusebius compares Constantine's achievements with those of Alexander, finding in favor of Constantine, of course.

The Metropolitan Museum of Art in New York City owns a colossal marble head of Constantine, c. three times life-size. Although it is without the diadem which we might have expected, it has been dated to 325/6. We must note a pronounced ridge on his nose and his disproportionately large eyes turned toward heaven.

The most famous likeness of Constantine is, without question, the colossal marble head in the cortile of the Palazzo dei Conservatori on Rome's Capitoline Hill. To have seen it once is to never forget it. It stands eight-and-a-half feet high and weighs eight to nine tons. The seated statue to which it once belonged was more than thirty feet tall, that is, seven to eight times life-size. In addition to the head there are a fragment of the torso, parts of both legs and feet, the right arm, and, strangely, two right hands, slightly different from each other. All these parts were found in 1487 in the Basilica Nova in the Forum and transferred to the Capitoline Hill by order of Pope Innocent VIII. In the head we note a strong chin, a nose ridge even more pronounced than on the New York head, and once again disproportionately large eyes turned upwards. The total impression is that of a fixed gaze, of imperturbability and of an almost supernatural majesty (see Figure 10).

Dating of this colossal head has posed problems. On the one hand there is good reason to assign it to the years 312–15, when Constantine was completing the Basilica Nova according to his own plans, adding another apse. Besides, Eusebius reports that Constantine had a statue of himself erected in Rome right after entering the city. On the other hand the stylistic features of the head preclude such an early date and suggest a date closer to 330. The ingenious solution which has been proposed is that the statue was originally erected in 312–15 and substantially reworked at some time after 325, especially to make the head conform to the new model. The presence of two right hands would thus also be explained: one, perhaps holding a scepter, had been replaced by the other, perhaps holding a cross-scepter. Two questions remain: are there any signs of reworking on the head, and why was the original right hand not destroyed when another was substituted for it? But *if* this theory is correct, then one may, perhaps, suppose further that the reworking of the statue was done in time for Constantine's visit to Rome in 326.

Another fine portrait of Constantine is to be seen in the National Museum of Belgrade. This is a life-size bronze head which hails from Naissus (Nish), Constantine's birthplace; it has been dated to *c*. 330. Again the impression is one of a powerful personality. Again we note an aquiline nose, but the upward glance is less remote. The emperor wears a jeweled diadem.

Two other marble statues of Constantine deserve mention, both dated *c*. 320. Both show their subject in military dress and originally were part of the Constantinian complex on Rome's Quirinal Hill (see Chapter 2). One now stands in the narthex of the Basilica of St John Lateran, the other on the balustrade of the Campidoglio.

Upon Constantine's death four different types of consecration coin were minted. One of these is rather remarkable. It is a small coin of the type known as *nummus centenionalis* and is of billon, an alloy of copper and silver. It was issued by several mints in both East and West after Constantine's sons had assumed the title of Augustus. On the obverse there is a veiled head of Constantine and the legend "The deified Constantine, father of the Augusti." Although other pagan iconographic features of consecration coins, such as the funeral pyre or the eagle, have been omitted, there is no break here with the pagan tradition: the emperor has become *divus*. On the reverse Constantine, dressed in a cloak, his right hand stretched out, is ascending to heaven on a quadriga, while the right hand of God (common enough in later Christian art) reaches out to meet him. While some Christians might have been reminded of the prophet

Figure 10 Colossal marble head of Constantine. Rome, Palazzo dei Conservatori.

Photograph: Hans A. Pohlsander.

Elijah's ascent to heaven on a chariot of fire, most people would more readily have seen an allusion to Sol/Helios/Apollo riding on his chariot of the sun. Solar symbolism thus was employed in Constantine's service to the end. Eusebius describes the coin, but without referring to its pagan connotations, of which he surely was aware.

13

An assessment

It remains for us now to assess Constantine's personality and his place in history. Unlike Nero and Domitian on the one hand and Antoninus Pius and Marcus Aurelius on the other, Constantine cannot simply be assigned to the list of the "bad" emperors or to that of the "good" emperors. Any such attempt would not do justice to the complexity of the record. That complexity is reflected in the fact that there is no consensus of scholarly opinion on important aspects of Constantine's person and reign. Not surprisingly, it is especially in the religious sphere that we note this lack of consensus. At one end of the spectrum of opinion, Alistair Kee concludes that Constantine's imperial ideology conquered the church and betrayed Christ. At the other end of the spectrum there is the judgment of Paul Keresztes, who holds that Constantine was "a truly great Christian Emperor and a genuine Apostle of the Christian Church."

No history of ancient Rome can end without a chapter on Constantine, and no history of Byzantium can begin without such a chapter. In this we may see a measure of the man's historical importance, of his position at a critical juncture in history. A history of the early Christian church that neglects to include him is equally unthinkable. And thus it is that both secular and ecclesiastical historians have accumulated a vast secondary literature on our subject, so vast indeed that no scholar can claim to control it.

Some scholars have spoken of a "Constantinian revolution," while others have avoided that term. Why is that so? It is true that in the course of his career Constantine made two epochal decisions: to support Christianity, and to establish a new capital in the East. These decisions, however, did not cause a break with the past in many aspects of the life of the empire. There was no radical reordering of society; neither the emperor nor the church seems to have aimed at such. Although Constantine issued a large number of legal decisions ("rescripts"), including some that were prompted by his Christian faith, he did not undertake a revision of the legal system in the light of Christian ethics. In government and in the army many traditional institutions continued to exist. "As an administrator, he was more concerned to preserve and modify the imperial system which he had inherited than to change it radically – except in one sphere," to quote Timothy D. Barnes. Christian art, having no substantial traditions of its own, could only adopt and adapt the norms and forms of traditional Greek and Roman art.

In continuing and completing the work of Diocletian, Constantine established the basis of the Byzantine state which was to exist for more than a millennium after him; this is no mean achievement. Ten emperors after him, including the last emperor of Byzantium, carried his name, and a new emperor could receive no greater compliment than to be called "a new Constantine." This Byzantine state derived a substantial part of its identity from its religion; it was built on an alliance of throne and altar. But the two partners in this alliance were not equals; there was always a preponderance of imperial authority over ecclesiastical authority. Although Constantine had been careful to let the bishops, assembled in council, make decisions in spiritual matters, he stands at the beginning of that order of church-state relationships which has long, but not very accurately, been called Caesaropapism. The modern view of that order is not usually a positive one.

In the end we must, with A. H. M. Jones, deny to Constantine the title of Great. We can do so without denying his excellence as a general and his historical importance. But we must find fault with his character and many of his deeds. Tempestuous by nature, he often made hasty decisions which he later regretted or made threats which, fortunately, he often did not fulfill. It is one of history's great ironies that the man named Constantine lacked constancy. His ecclesiastical policy in the years 325–37 shows this especially well. Key players, such as Athanasius, Arius and Eusebius of Nicomedia, enjoyed his favor or fell out of favor, sometimes with astonishing rapidity. Ramsay MacMullen speaks of Constantine's "tergiversations."

The arrangements which Constantine made, or failed to make, for his own succession show him at his weakest, for they are entirely unsatisfactory. He had spent the most productive years of his life striving for sole power and ultimately achieving it. Augustus, Vespasian, Trajan, Hadrian, and Antoninus Pius could have taught him how to pass it on to a successor. Diocletian's noble, if less successful, efforts in this area should have taught him additional lessons. And it cannot be said that fate granted him insufficient time. Ultimately the blame for the bloody coup of 337 falls squarely on the shoulders of the man who was in the best position to forestall it, Constantine himself.

Even the founding of Constantinople is less than an unqualified success story. It is the old Rome on the Tiber, not the New Rome on the Bosporus, which remained central to European history through the centuries. And is it possible that the Latin church of the West and the Greek church of the East would have remained one if Constantine had not moved the capital of the empire to the East? And it may be argued that the centralization of imperial wealth and power in the East weakened the West and hastened its eventual collapse.

In the Orthodox churches Constantine is recognized as a saint, sharing a feast day, 21 May, with his mother and additionally having a feast day of his own on 3 September. From the fifth century on, beginning with the historian Theodoret, it became fashionable in the East to call Constantine *isapostolos* (equal to the apostles). He holds that rank to this day in the liturgy of the Greek Orthodox Church, where he is specifically compared to the Apostle Paul. But we must deny to Constantine the title of Saint and take the risk of offending Orthodox believers. We need not deny that he committed himself to the Christian religion, that he thought of himself as God's servant and that he meant to lead all men to the true worship of the supreme deity. We can excuse the late date of his baptism.

But Constantine died with blood on his hands. We need not

count against him the thousands who died in his wars; in waging these wars he did what probably any other man in the same position would have done. He brought the blessings of peace by the horrors of war. More attention has been paid to the fact that Constantine was in various ways and to various degrees responsible for the deaths of so many of his nearest kin. Let us not indict him in the case of Maximian; the old man may have "asked for it." Maxentius died in battle, but nothing at all suggests that his life would have been spared had he been taken alive. Let us withhold judgment in the case of Crispus and Fausta, since we do not have sufficient knowledge of all the facts and circumstances. But certainly Constantine stands guilty of the murder of the Licinii, father and son.

Certain key concepts of the Christian faith, such as repentance, atonement and redemption, Constantine never made his own. In the words of one scholar (Joseph Vogt), Constantine "remained alien to the greater depths of Christian belief." In the words of another (Ramsay MacMullen), "inwardness was something on which he never wasted much time." A third (Alistair Kee) observes that Constantine's religion was "neither profound nor particularly edifying," although he was "fanatically committed to it." There was not enough of a spiritual quest and too much concern with external and material matters: buildings, endowments, honors, and privileges.

Let us grant that humility was not cherished in the Roman system of values and is hard to achieve for those who are in positions of great power. Nevertheless it is a quality which we have a right to expect in a Christian saint. And Constantine was utterly without it. The arrangements which he made for his own burial were presumptuous, if not blasphemous. His church foundations were not only expressions of piety; they were also celebrations of his victories. The inscription which he caused to be placed on the triumphal arch of Old St Peter's in Rome (see Chapter 5) is revealing. It read: "Because under your leadership the world rose triumphantly to the heavens Constantine Victor has dedicated this building to you." Eusebius reports that Constantine built the basilica at Nicomedia to celebrate his own victory and the victory of his Savior. And Eusebius saw nothing reprehensible in Constantine's likening his own achievement to that of Christ.

Let us be content with recognizing in Constantine a resolute ruler, a superb general who never lost a battle and "the most tireless worker for Christian unity since St Paul," to quote Robin Lane Fox. It is personalities of such strengths and such shortcomings who shape history, capture our imagination and hold our attention. Bernini's magnificent equestrian statue of Constantine on the Scala Regia of the Vatican is exciting to us not only because of its artistic excellence but also because of the man whom it represents.

APPENDIX I

The sources for the reign of Constantine

The literary sources for the reign of Constantine are neither as complete nor as unbiased as we might wish. Information on the secular aspects of his reign is particularly inadequate.

Among the literary sources, the writings of Eusebius of Caesarea (*c.* 260–339) are the most important. Eusebius' *History of the Church* was published in its fourth and final edition *c.* 325, after the fall of Licinius in 324, but before the death of Crispus in 326. It has rightly been called a massive achievement and earned its author the title "father of church history." On the occasion of Constantine's *tricennalia* in 336 Eusebius delivered his oration *In Praise of Constantine; On Christ's Sepulchre*, a description of the Church of the Holy Sepulchre in Jerusalem, produced in the previous year, is appended to this. *The Life of Constantine*, written after the emperor's death in 337 and showing signs of being unfinished, is an encomium rather than a true biography. It is now generally recognized as the work of Eusebius.

Eusebius' work is not without its shortcomings: his admiration of Constantine knew no bounds, and he is not above suppressing, distorting or misrepresenting facts to achieve his purpose. But Jacob Burckhardt surely judges him too severely when he calls him "the first thoroughly dishonest historian of antiquity." We should be grateful to him for his practice of quoting, in full, the text of numerous Constantinian documents, such as letters, decrees and speeches.

Rufinus of Aquileia (*c.* 345–410) translated Eusebius' *History of the Church* into Latin and added to the original ten books two more of his own, carrying the story to the year 395. In the fifth century Socrates (Scholasticus), Sozomen and Theodoret wrote histories of the church which overlap with and sometimes supplement or correct the final portions of Eusebius' *History of the Church* and then provide their own account of the years 324–37 (and beyond).

Among the twelve Latin *Panegyrics* there are five (nos 4–7 and 12 in the editions of Baehrens and Mynors; nos 6–10 in the edition of Galletier) which are addressed to Constantine; these were delivered by Gallic orators in the years 307–21. They are highly tendentious and offend modern taste by their flowery style and fulsome flattery. Nevertheless they provide some insight into their times and some details on the earlier years of Constantine's reign.

Lactantius (*c.* 240–*c.* 320) was a native of North Africa, was appointed professor of rhetoric at Nicomedia and for some years served as tutor of the young Crispus. His most significant theological work is the *Divine Institutions*, but the student of the age of Constantine will be more interested in a shorter treatise, *On the Deaths of the Persecutors*, published probably in 314. The point of this treatise is that all the persecutors have come to a bad end – the agonies of Galerius are described in grisly detail – but also that good, in the person of Constantine, prevails over evil.

Constantine's own *Oration to the Assembly of the Saints*, although its authorship has been questioned and its date is controversial, allows us to assess his religious–political ideology. It was recorded for us by Eusebius as an appendix to the *Life of Constantine*.

The *Origo Constantini Imperatoris* is a concise but rich biography which can be dated to the fourth century; it attains a high degree of reliability and objectivity, often providing information not given by Eusebius. It constitutes the first part of a manuscript known as the *Anonymus Valesianus* or *Excerpta Valesiani* (after the French scholar Henri de Valois, who first edited it in the seventeenth century). The most useful modern edition is that of Ingemar König (Trier 1987), with German translation and historical commentary.

The *Historia Nova* of Zosimus (dating from the early sixth century) was written from a pagan perspective and is exceedingly hostile to Constantine.

It is very much to be regretted that we have lost the first thirteen books of the *History* of Ammianus Marcellinus (*c.* 330–95). Intended as a continuation of Tacitus, this *History* covered the years 96–378; the lost books would have given us, among other things, an account of the reign of Constantine.

Constantine's legal decisions, numbering in the hundreds, can be found in the *Codex Theodosianus*, the codification of Roman law undertaken on the initiative of the emperor Theodosius II (408–50). Unfortunately the usefulness of the collection is somewhat diminished by the fact that the subscriptions given with each legal decision are notoriously unreliable, both as to location and as to date.

An excellent selection of Latin inscriptions relating to Constantine is provided by Hermann Dessau, *Inscriptiones Latinae Selectae* (Berlin 1892–1916), vol. I.

The coinage of the reign of Constantine can be bewildering because of the great variety of different coin types that were produced. Some Constantinian coins are of unusual documentary value; many illustrate the religious change in the empire.

For primary source materials available in English, readers should consult the bibliography.

APPENDIX II

Glossary of Greek, Latin and technical terms

Canon A regulation or dogma decreed by a church council.

Chi-Rho A monogram consisting of the Greek letters X (Chi) and P (Rho) and signifying Christ (ΧΡΙΣΤΟΣ).

Consistorium A group of men "standing together," the emperor's council of advisors.

Consubstantialis See **homoousios**.

Consularis A provincial governor of consular rank.

Curiales The members of local city councils, also known as *decuriones*, and their descendants.

Damnatio memoriae The erasure of a person's name from public records and monuments. The Roman senate decreed *damnatio memoriae* for several "bad" emperors.

Decennalia A festival celebrated every ten years, specifically during the tenth year of an emperor's reign.

Dies imperii The anniversary day of an emperor's gaining *imperium*, that is, of his accession.

Domus ecclesiae A private house used as place of Christian assembly and worship. Good examples are the house-church under the Church of S Clemente in Rome and the house-church in Dura-Europos on the Euphrates.

Filius Augusti "Son of Augustus," a title bestowed by Galerius on Maximinus Daia and Constantine to placate them.

Homoousios From Greek *homos* = equal, same, and *ousia* =

substance, essence. Latin *consubstantialis*; English "consubstantial." At the Council of Nicaea Constantine proposed this term to define the relationship of the Son to the Father.

Hypostasis Greek for "substance" or "essence;" also *ousia*; Latin *substantia*.

Labarum A *vexillum* or military standard crowned by the Chi-Rho.

Liber Pontificalis *The Book of the Popes*, a collection of papal biographies. It was first compiled in the sixth century; it terminates with Pope Pius II (d. 1464).

Manichaeism The religious movement founded by the Persian prophet Mani in the third century. It incorporates elements of Judaism, Christianity, Buddhism, and Zoroastrianism and is marked by dualism.

Nobilissimus/nobilissima A title granted during the later empire to members of the imperial family.

Ousia See **hypostasis**.

Panegyric A formal speech of praise, following certain rules of rhetoric. Panegyric as a literary genre was developed by the Greeks; Isocrates composed a panegyric for Philip II of Macedon. Pliny the Younger wrote a panegyric for the emperor Trajan. A collection of twelve Latin panegyrics includes five for Constantine.

Pax deorum "Peace with the gods," the harmonious relationship between the Roman people and the gods.

Pontifex Maximus The highest-ranking religious official of the Roman government. Beginning with Augustus the emperors claimed this post for themselves.

Porta Nigra The "Black Gate," the north gate of the walls of Trier, built in the second century, the most famous landmark of the city to this day.

Praeses "Presiding officer," a provincial governor ranking below a *consularis*, a provincial governor of consular rank.

Praetorian Guard An elite military force founded by Augustus and stationed in Rome. It often played a decisive but unwelcome role in choosing an emperor. It was disbanded by Constantine in 312.

Praetorian Prefect Originally the Commander of the Praetorian Guard. In the later empire there could be several praetorian prefects, representing the emperor in major divisions of the empire.

Princeps The title preferred by Augustus to describe his position in the state; hence the "principate."

Sassanid or Sassanian This term refers to the Persian empire and its rulers 224–636; to be distinguished from Achaemenid Persia.

Secular Games Games and sacrifices held to mark the end of a *saeculum*. The Secular Games held by Augustus in 17 BC and those held by Philip the Arab in 248 are best known.

Sibylline Books The collection of the prophecies of the Cumaean Sibyl, the priestess of Apollo at Cumae. These books were kept in the Capitol by a college of fifteen priests and were consulted only by order of the senate.

Substantia See **hypostasis**.

Tetrarchy Government by four. The First Tetrarchy consisted of Diocletian, Maximian, Galerius, and Constantius; the Second Tetrarchy of Galerius, Constantius, Severus, and Maximinus Daia.

Traditor A Christian who, at a time of persecution, surrendered the Scriptures to the authorities; hence English "traitor."

Tribunes High-ranking military officers; originally there were six assigned to each legion.

Tricennalia A festival celebrated every thirty years, specifically during the thirtieth year of an emperor's reign.

Vicarius In the territorial organization instituted by Diocletian a *vicarius* ("personal representative") governed a diocese, which normally consisted of three provinces.

Vicennalia A festival celebrated every twenty years, specifically during the twentieth year of an emperor's reign.

APPENDIX III

Biographical notes

Agritius Bishop of Trier in the early fourth century. He is known to have attended the Council of Arles in 314 and apparently had close ties to the empress Helena during her residence in Trier.

Alexander Bishop of Alexandria *c*. 312–28. He had to contend with both the Melitian schism and the Arian controversy; he attended the Council of Nicaea in 325.

Ambrose Bishop of Milan 374–97. He was a firm defender of Catholic orthodoxy against both Arians and pagans, promoted the cult of the martyrs and was instrumental in the conversion of St Augustine. In two confrontations he stood his ground against the emperor Theodosius I.

Antoninus Pius Emperor 138–61. One of the adoptive emperors of the second century, he had been adopted by his predecessor Hadrian, while he in turn adopted Marcus Aurelius. He, Marcus Aurelius and Commodus are collectively known as the Antonines.

Arius Presbyter in Alexandria. He taught that the Son is not co-eternal with and not equal to the Father. He was condemned by a local synod in 318 or 319 and by the Council of Nicaea in 325. His strongest supporter was Eusebius of Nicomedia. He died in 336 at Constantinople.

Athanasius Deacon in Alexandria under Bishop Alexander and bishop of Alexandria 328–73, although repeatedly exiled. He was adamantly opposed to the readmission of Arius and often at odds with the Arian emperor Constantius II.

Augustine, Saint 354–430, Bishop of Hippo in North Africa and "Doctor of the Church;" author of *City of God* and *Confessions.*

Augustus (Octavian) Emperor 27 BC – AD 14. He provided for the trouble-free succession of his stepson and adoptive son Tiberius.

Aurelian Emperor 270–75. He built the great walls of Rome which bear his name. He celebrated a triumph over Queen Zenobia of Palmyra and put down a separatist movement in Gaul. He was devoted to the sun god *Sol Invictus.*

Aurelius Victor Fourth-century historian and author of *Caesares,* a collection of imperial biographies. Governor of Pannonia in 361 and prefect of Rome in 389.

Carausius Roman naval commander in the North Sea. He seized Britain and parts of Gaul and declared himself emperor. He was murdered in 293 while under attack by Constantius I.

Claudius II Gothicus Emperor 268–70. He won a splendid victory over the Goths at Naissus (Nish); hence his name. Constantine was to claim him as an ancestor.

Constans Youngest of the four sons of Constantine I. He was proclaimed Caesar in 333 and was Augustus in 337–50. Controlling initially only Italy, Africa, and Illyricum, he became ruler of the entire West after defeating and killing his elder brother Constantine II in 340.

Constantia Half-sister of Constantine I. She was married to the emperor Licinius in 313. After the latter's death she occupied a position of honor at Constantine's court. She was an Arian sympathizer.

Constantine II Oldest of the three sons born to Constantine by Fausta. He was proclaimed Caesar in 317 and was Augustus in 337–40. He was defeated and killed by his younger brother Constans when he invaded Italy.

Constantius I Father of Constantine I, Caesar 293–305 and Augustus 305–6. He resided, mostly, at Trier, but died at York. During the Great Persecution he gave only token compliance to Diocletian's edicts.

Constantius II Second oldest of the three sons born to Constantine by Fausta. He was proclaimed Caesar in 324 and was

95

Augustus in 337–61, ruling both East and West from 350. In the Arian–Athanasian dispute he supported the Arian party.

Crispus Oldest of the four sons of Constantine I, born to him by Minervina *c.* 305. He was proclaimed Caesar in 317 and served with distinction in the war against Licinius, but was put to death, on his father's orders, in 326 at Pola in circumstances which are not entirely clear.

Decius Emperor 249–51. In 250 he launched the first empire-wide and systematic persecution of the Christians. He perished in battle with the Goths.

Eusebius of Caesarea Bishop of Caesarea in Palestine *c.* 313–39. He attended the councils of Antioch (325), Nicaea (325) and Tyre (335). He has been called the "father of church history;" see Appendix I.

Eusebius of Nicomedia Bishop, 317–42, first of Berytus (Beirut), then of Nicomedia, and finally of Constantinople. He was a defender of the person and doctrines of Arius. Deposed in 325, he was restored two years later, became Constantine's chief ecclesiastical advisor, and baptized him in 337.

Eustathius Bishop of Antioch in Syria. He was a staunch defender of Athanasian orthodoxy, but was deposed in 327, possibly after having run foul of the empress Helena.

Eutropia Wife of the emperor Maximian and mother of Maxentius and Fausta. Constantine apparently held her in high esteem, even after the death of Fausta; she visited the Holy Land and reported to him on the state of the sanctuary at Mamre.

Fausta Daughter of the emperor Maximian. She was married to Constantine in 307, bore him five children and was proclaimed Augusta after the victory over Licinius. She was put to death, on Constantine's orders, in 326 in circumstances which are not entirely clear.

Gallienus Emperor 260–8. He put an end to the persecution of the Christians which had been launched by his father Valerian.

Gregory the Illuminator, Saint "The Apostle to the Armenians." He converted King Tiridates III of Armenia, early in the fourth century, and became the first Catholicos of the Armenian Church.

Hadrian Emperor 117–38. One of the adoptive emperors of the second century, he had been adopted by his predecessor Trajan, while he in turn adopted Antoninus Pius.

Helena Mother of Constantine I. She was of humble origins but rose to the rank of Augusta. She undertook a pilgrimage to the Holy Land, probably in 326. Tradition attributes to her the discovery of the True Cross, and she became a saint in both the Western and the Eastern church.

Jerome, Saint (Hieronymus) *c*. 347–419/20, scholar, translator, and ascetic. He translated the *Chronicle* of Eusebius of Caesarea from Greek into Latin and carried it forward to 378. He gave us the Vulgate version of the Bible. He attended the Council of Constantinople in 381.

Julian the Apostate Emperor 361–63. He was the son of Julius Constantius, one of the half-brothers of Constantine I. His attempts to roll back the advance of Christianity were cut short by his death on a campaign against the Persians.

Justinian Emperor 527–65. He rebuilt several churches originally founded by Constantine: at Bethlehem the Church of the Nativity and at Constantinople the churches of Hagia Sophia and Hagia Eirene.

Lactantius 245/50–325, Christian scholar; see Appendix 1.

Licinius II Son of Licinius I and Constantia. He was proclaimed Caesar in 317, but put to death by Constantine in the spring of 325.

Lucian, Saint Head of a theological school at Antioch, where his students included Arius and Eusebius of Nicomedia. He was martyred in 312 at Nicomedia and became the favorite saint of Helena.

Lucius Verus Emperor 161–9; co-emperor with Marcus Aurelius.

Macarius Bishop of Jerusalem at the time of Helena's pilgrimage. In a letter preserved by Eusebius, Constantine directed him to build the Church of the Holy Sepulchre.

Marcus Aurelius Emperor 161–80, with Lucius Verus as co-emperor 161–69. He designated his son Commodus as successor.

Miltiades Pope 311–14. In 313 he assembled a council of bishops at Rome to deal with the Donatist problem, but he did not attend the Council of Arles in the following year.

Narses King of Persia 293–302. He was decisively defeated by

97

the emperor Galerius in 298; the event is commemorated by the Arch of Galerius in Thessalonike.

Origen *c.* 185–*c.* 251, Christian scholar. He was a native of Alexandria and at an early age became head of the catechetical school of Alexandria, but later was active at Caesarea. He was a prolific writer and exerted a profound influence on Christian thought.

Ossius (Hosius) *c.* 257–*c.*357, bishop of Cordoba, Spain. As chief ecclesiastical advisor to Constantine for some years, 312–26, he played an important role before and during the Council of Nicaea.

Philip the Arab Emperor 244–9. In 248, observing the 1000th anniversary of the city of Rome, he held the last Secular Games.

Philostorgius *c.* 368–*c.* 439, church historian. He was of the Arian persuasion, and his work is extant only in fragments.

Septimius Severus Emperor 193–211 and founder of the Severan dynasty. His triumphal arch stands in the Forum Romanum to this day. He undertook considerable building activity at Byzantium.

Shapur I King of Persia 241–72. In 260 he defeated and captured the emperor Valerian.

Shapur II King of Persia 309/10–379. Constantine wrote to him on behalf of the Christians in Persia and, at the time of his death, was preparing a campaign against him.

Socrates (Scholasticus) Fifth-century church historian; see Appendix I.

Sozomen Fifth-century church historian; see Appendix I.

Sylvester Pope 314–35. An erroneous but persistent tradition holds that he baptized the emperor Constantine and that he received the Donation of Constantine. He did not attend the Council of Nicaea.

Theodora Daughter or stepdaughter of the emperor Maximian. No later than 293 she married Constantius I, who was required to put aside Helena at that time. She bore him six children.

Theodoret Fifth-century church historian; see Appendix I.

Theodosius II Emperor 408–50. He expanded the city of Constantinople, building new walls, which stand to this day. A great achievement of his reign was the publication, in 438, of the *Codex Theodosianus*, a compilation of Roman laws and a valuable source of information on the reign of Constantine; see Appendix I.

Tiridates III King of Armenia *c.* 287–? He was on friendly terms with the Romans and was converted to Christianity, early in the fourth century, by Gregory the Illuminator.

Trajan Emperor 98–117. One of the adoptive emperors of the second century, he had been adopted by his predecessor Nerva, while he in turn adopted Hadrian.

Valerian Emperor 253–60. In 258 he resumed the persecution of the Christians. He was defeated and captured by the Persian king Shapur I in 260. He suffered much humiliation before finally being put to death.

Vespasian Emperor 69–79 and founder of the Flavian dynasty. He provided for the trouble-free succession of his son Titus.

Zosimus Early sixth-century pagan (!) historian; see Appendix I.

APPENDIX IV

The Creeds

1 The creed introduced by Eusebius of Caesarea at the Council of Nicaea:

We believe in one God, the Father, the Almighty, maker of all things both visible and invisible. And in one Lord Jesus Christ, the Word of God, God from God, light from light, life from life, the only-begotten Son, first-begotten of all creation, begotten before all ages from the Father, through Whom also all things came into being, Who because of our salvation became flesh, and dwelt among men, and suffered, and rose again on the third day, and ascended to the Father, and will come again in glory to judge the living and dead. We believe also in one Holy Spirit.

(Socrates, *History of the Church* 1.8.3; author's translation)

2 The creed adopted by the Council of Nicaea:

We believe in one God, the Father, the Almighty, maker of all things both visible and invisible. And in one Lord Jesus Christ, the Son of God, begotten from the Father, only-begotten, that is, from the substance of the Father, God from God, light from light, true God from true God, begotten not made, of the same substance with the Father, through Whom all things came into being, both those in heaven and those on earth, Who because of us men and because of our salvation came down and became flesh, became man, suffered, and rose again on the third day,

ascended to the heavens, coming to judge the living and the dead. And in the Holy Spirit. But as for those who say, that there was when He was not, and that before He was born He was not, and that He came into existence out of nothing, or who say that He is of a different hypostasis or substance, or that the Son of God is subject to alteration or change, these the catholic and apostolic Church anathematizes.
(Socrates, *History of the Church* 1.8.4; author's translation)

3 The Nicaeno–Constantinopolitan Creed, commonly called the "Nicene Creed":

We believe in one God, the Father, the Almighty, maker of heaven and earth, of all things both visible and invisible. And in one Lord Jesus Christ, the only-begotten Son of God, who was begotten from the Father before all ages, light from light, true God from true God, begotten not made, of the same substance with the Father, through Whom all things came into existence, Who because of us men and because of our salvation came down from heaven, and became flesh from the Holy Spirit and the Virgin Mary and became man, and was crucified for us under Pontius Pilate, and suffered and was buried, and rose again on the third day according to the Scriptures, and ascended to heaven, and sits at the right hand of the Father, and is coming again with glory to judge the living and the dead, of Whose kingdom there will be no end. And in the Holy Spirit, the Lord and life-giver, Who proceeds from the Father,* Who together with the Father and the Son is worshipped and glorified, Who spoke through the prophets. In one holy catholic and apostolic Church. We confess one baptism for the remission of sins. We look forward to the resurrection of the dead and the life of the age to come. Amen.
(*Acta Conciliorum Oecumenicorum*, ed. Eduard Schwartz (Berlin and Leipzig 1914 ff.) vol. II. 1.2, p. 79 f; author's translation)

* At this point in the creed the Western church, first in Spain and then elsewhere, added the clause "and from the Son" (*filioque*). This is the basis of the doctrine of "double procession." It was never accepted in the East and was one of the differences between the Western church and the Eastern church that ultimately led to the great schism of 1054.

Bibliography

Primary source material in English

P. M. Bruun, *The Roman Imperial Coinage*, vol. VII: *Constantine and Licinius*, AD 313–337 (London: Spink & Son, 1966).

P. R. Coleman-Norton, *Roman State and Christian Church*, vol. I (London: SPCK, 1966).

R. Davis, *The Book of the Pontiffs (Liber Pontificalis)*, Translated *Texts for Historians*, Latin Series V (Liverpool: Liverpool University Press, 1989).

Eusebius, *The History of the Church*, translated by G. A. Williamson, revised and edited with a new introduction by Andrew Louth (London: Penguin, 1989).

Eusebius, *The Life of Constantine*, translated by C. McGiffert and E. C. Richardson, in *Select Library of Nicene and Post-Nicene Fathers*, 2nd series, vol. I (New York: Christian Literature, 1890; reprinted Grand Rapids, MI: William B. Eerdmans, 1978).

Eusebius, *Panegyric to Constantine*, in H. A. Drake, *In Praise of Constantine, A Historical Study and New Translation of Eusebius' Tricennial Orations* (Berkeley: University of California Press, 1976).

Lactantius, *De Mortibus Persecutorum (On the Deaths of the Persecutors)*, edited and translated by J. L. Creed (Oxford: Clarendon Press, 1984).

L. R. Loomis, *The Book of the Popes* (New York: Columbia University Press, 1916; reprinted 1965).

Origo Constantini Imperatoris or *Anonymus Valesianus*, translated by J. C. Rolfe, in Ammianus Marcellinus, vol. III, *Loeb Classical Library* (Cambridge, MA: Harvard University Press, 1952), 506–31.

C. Pharr *et al.*, *The Theodosian Code and the Sirmondian Constitutions* (Princeton: Princeton University Press, 1952).

102

J. Stevenson, *A New Eusebius: Documents Illustrating the History of the Church to* AD *337*, revised edn (London: SPCK, 1987).

C. H. V. Sutherland, *The Roman Imperial Coinage, vol. VI: From Diocletian's Reform (*AD *294) to the Death of Maximinus (*AD *313*) (London: Spink & Son, 1967).

Secondary literature in English

The emperor Constantine and especially his conversion to Christianity have prompted a vast amount of secondary literature. Only a small selection of titles in English is listed here.

The two best introductions to the general subject are:

A. H. M. Jones, *Constantine and the Conversion of Europe*, (London: English Universities Press, 1948 and New York: Macmillan, 1962).

R. MacMullen, *Constantine* (New York: The Dial Press, 1969).

More advanced or more specialized studies are:

S. S. Alexander, "Studies in Constantinian Church Architecture," *Rivista di archeologia cristiana* 47 (1971) 281–330 and 49 (1973) 33–44.

A. Alföldi, *The Conversion of Constantine and Pagan Rome* (Oxford: Clarendon Press, 1948 and 1969).

G. T. Armstrong, "Imperial Church Building in the Holy Land in the Fourth Century," *Biblical Archaeologist* 30 (1967) 90–102.

L. W. Barnard, "Church-State Relations, AD 313–337," *Journal of Church and State* 24 (1982) 337–55.

T. D. Barnes, "Lactantius and Constantine," *Journal of Roman Studies* 63 (1973) 29–46.

T. D. Barnes, "The Emperor Constantine's Good Friday Sermon," *Journal of Theological Studies, new series* 27 (1976) 414–23.

T. D. Barnes, *Constantine and Eusebius* (Cambridge, MA: Harvard University Press 1981).

T. D. Barnes, *The New Empire of Diocletian and Constantine* (Cambridge, MA: Harvard University Press, 1982).

T. D. Barnes, *From Eusebius to Augustine*, Part Two (Aldershot, Hampshire: Variorum, 1994).

N. H. Baynes, *Constantine the Great and the Christian Church* (London: British Academy, 1931, reissued New York: Oxford University Press, 1975).

R. Beny and P. Gunn, *The Churches of Rome* (New York: Simon & Schuster, 1981).

G. C. Brauer, Jr., *The Age of the Soldier Emperors: Imperial Rome,* AD *244–284*, (Park Ridge, NJ: Noyes Press, 1975).

R. A. G. Carson, *Coins of the Roman Empire* (London and New York: Routledge, 1990).

H. Chadwick, "Conversion in Constantine the Great," *Studies in Church History* 15 (1978) 1–13.

V. C. De Clercq, *Ossius of Corduba: A Contribution to the History of the Constantinian Period* (Washington, DC: Catholic University of America Press, 1954).

J. W. Crowfoot, *Early Churches in Palestine* (London 1941; reprinted College Park, MD: University of Maryland Press, 1971).

L. D. Davis, *The First Seven Ecumenical Councils (325–787): Their History and Theology* (Collegeville, MN: Liturgical Press, 1990), Chapter 1–3.

J. W. Drijvers, "Flavia Maxima Fausta: Some Remarks," *Historia* 41 (1992) 500–6.

J. W. Drijvers, *Helena Augusta: The Mother of Constantine the Great and the Legend of Her Finding the True Cross* (Leiden: E. J. Brill, 1992).

W. H. C. Frend, *The Donatist Church* (Oxford: Clarendon Press, 1952, 1976 and 1985).

W. H. C. Frend, *The Rise of Christianity* (Philadelphia: Fortress Press, 1984), Chapter 13–15.

Edward Gibbon, *The History of the Decline and Fall of the Roman Empire*, ed. J. B. Bury, 7 vols (London: Methuen, 1896–1900).

J. Harries and I. Wood (eds), *The Theodosian Code* (Ithaca, NY: Cornell University Press, 1993), Part II.

E. D. Hunt, *Holy Land Pilgrimage in the Later Roman Empire AD 312–460* (Oxford: Clarendon Press, 1982), Chapters 1–2.

A. H. M. Jones, *The Later Roman Empire, 284–602*, 2 vols (Oxford: Blackwell, 1964; reprinted Baltimore: The Johns Hopkins University Press, 1986).

A. Kee, *Constantine Versus Christ* (London: SCM Press, 1982).

P. Keresztes, *Constantine: A Great Christian Monarch and Apostle*, (Amsterdam: J. C. Gieben, 1981).

R. Krautheimer, "The Constantinian Basilica," *Dumbarton Oaks Papers* 21 (1967) 115–40.

R. Krautheimer, *Rome: Profile of a City, 312–1308* (Princeton: Princeton University Press, 1980), Chapter 1.

R. Krautheimer, *Three Christian Capitals: Topography and Politics* (Berkeley: University of California Press, 1983).

R. Krautheimer, *Early Christian and Byzantine Architecture*, 4th edn (Harmondsworth, Middlesex: Penguin, 1986).

R. Lane Fox, *Pagans and Christians* (New York: Alfred A. Knopf, 1987), Chapter 12.

M. Maclagan, *The City of Constantinople* (New York: Praeger Publishers, 1968).

R. MacMullen, *Christianizing the Roman Empire* (New Haven: Yale University Press, 1984).

T. F. Mathews, *The Early Churches of Constantinople: Architecture and Liturgy* (University Park, PA: Pennsylvania State University Press, 1971).

R. L. Milburn, *Early Christian Art and Architecture* (Berkeley: University of California Press, 1988).

A. Momigliano (ed.), *The Conflict between Paganism and Christianity in the Fourth Century* (Oxford and New York: Oxford University Press, 1963).

C. Odahl, "The Celestial Sign on Constantine's Shields at the Battle of the Milvian Bridge," *Journal of the Rocky Mountain Medieval and Renaissance Association* 2 (1981) 15–28.

C. Odahl, "Constantine's Epistle to the Bishops at the Council of Arles: A Defence of Imperial Authorship," *The Journal of Religious History* 17 (1993) 274–89.

C. Odahl, "The Christian Basilicas of Constantinian Rome," *Ancient World* 26 (1995) 3–28.

J. Pelikan, *The Christian Tradition, vol. I: The Emergence of the Catholic Tradition (100–600)* (Chicago: University of Chicago Press, 1971), Chapter 4.

J. Pelikan, *The Excellent Empire: The Fall of Rome and the Triumph of the Church* (San Francisco: Harper & Row, 1987).

H. A. Pohlsander, "Crispus: Brilliant Career and Tragic End," *Historia* 33 (1984) 79–106.

H. A. Pohlsander, "Constantia," *Ancient Society* 24 (1993) 151–67.

H. A. Pohlsander, "The Date of the *Bellum Cibalense*: A Re-examination," *Ancient World* 26 (1995) 89–101.

R. H. Smith, "The Tomb of Jesus," *Biblical Archaeologist* 30 (1967) 74–90.

Y. Tsafrir (ed.), *Ancient Churches Revealed* (Jerusalem: Israel Exploration Society and Washington DC: Biblical Archaeology Society, 1993).

J. Vogt, *The Decline of Rome: The Metamorphosis of Ancient Civilization* (New York: Praeger, 1967), Chapter 2.

W. F. Volbach, *Early Christian Art* (New York: Abrams, 1962).

E. M. Wightman, *Roman Trier and the Treveri* (London: Praeger, 1970).

R. D. Williams, *Arius: Heresy and Tradition* (London: Darton, Longman and Todd, 1987).

S. Williams, *Diocletian and the Roman Recovery* (New York: Methuen, 1985).